23rd Autumn School of Geodesy 2017

E3S Web of Conferences Volume 55 (2018)

Walbrzych, Poland
21 – 22 September 2017

Editors:

Jan B. Blachowski
J. Gorniak-Zimroz
D. Kasza

ISBN: 978-1-5108-7181-6

Printed from e-media with permission by:

Curran Associates, Inc.
57 Morehouse Lane
Red Hook, NY 12571

Some format issues inherent in the e-media version may also appear in this print version.

Printed by Curran Associates, Inc. (2018)

For additional information, please contact EDP Sciences – Web of Conferences at the address below.

EDP Sciences – Web of Conferences
17, Avenue du Hoggar
Parc d'Activité de Courtabœuf
BP 112
F-91944 Les Ulis Cedex A
France

Phone: +33 (0) 1 69 18 75 75
Fax: +33 (0) 1 69 28 84 91

contact-edps@webofconferences.org

Additional copies of this publication are available from:

Curran Associates, Inc.
57 Morehouse Lane
Red Hook, NY 12571 USA
Phone: 845-758-0400
Fax: 845-758-2633
Email: curran@proceedings.com
Web: www.proceedings.com

TABLE OF CONTENTS

APPLICATION OF OBSERVATIONS OF RECENT TECTONIC ACTIVITY IN THE SWIEBODZICE DEPRESSION (THE SUDETES, SW POLAND) IN ASSESSING SEISMIC HAZARD IN THE FORE-SUDETIC MONOCLINE 1
Kaczorowski Marek, Kasza Damian, Zdunek Ryszard, Wronowski Roman

ENHANCEMENT OF THE WALBRZYCH HARD COAL MINES GEOGRAPHIC INFORMATION SYSTEM FOR ITS APPLICATION IN STUDIES OF MINING DEFORMATIONS 10
Blachowski Jan, Herkt Paulina

A REVIEW OF METHODS OF CARTOGRAPHIC PRESENTATION OF URBAN SPACE 17
Bac-Bronowicz Joanna, Wojciechowska Gabriela

DETECTION OF SURFACE SUBSIDENCE USING SAR SENTINEL 1A IMAGERY AND THE DINSAR METHOD – A CASE STUDY OF THE BELCHATOW OPEN PIT MINE, CENTRAL POLAND 25
Wajs Jaroslaw, Milczarek Wojciech J.

ANALYSIS OF TECTONIC ACTIVITY OF THE NORTH ANATOLIAN FAULT BASED ON SBINSAR METHOD 31
Owczarz Karolina

LOS DISPLACEMENTS OF MAUNA LOA VOLCANO, HAWAII ISLAND, AS DETERMINED USING SBAS-INSAR 38
Grzesiak Karolina, Milczarek Wojciech J.

CHANGES IN THE CONDUCT OF ADMINISTRATIVE PROCEEDINGS IN THE SCOPE OF CADASTER DATABASE 47
Sudo Ewa

APPLICATION OF NEW MEASUREMENT TECHNOLOGY FOR DEFORMATION STUDY OF STRUCTURES IN MINING AREAS 51
Gruchlik Piotr, Kowalski Andrzej

APPROXIMATION OF THE PROCESS OF CHANGES IN DEFORMATION OF LAND SURFACE USING ARTIFICIAL NEURAL NETWORKS 61
Mrówczynska Maria, Sztubecki Jacek

STABILITY TEST OF TCRP1201+ TOTAL STATION PARAMETERS AND ITS SETUP 66
Odziemczyk Waldemar

PROPOSITION OF DETERMINATION OF DISPLACEMENTS USING THE TDRA 6000 LASER STATION 76
Sztubecki Jacek, Mrówczynska Maria, Bujarkiewicz Adam

EXAMINATION OF VERTICAL DISPLACEMENTS AT NON-PERMANENTLY FIXED POINTS 81
Gradka Robert

ANALYSIS OF POWER LINES SPAN GEOMETRY BASED ON TLS MEASUREMENTS 88
Kwinta Andrzej, Wazydrag Karolina, Zygmunt Mariusz

MODELLING OF UNDERGROUND OBJECTS AND GEOLOGICAL-TECTONIC STRUCTURES FROM A POINT CLOUD: APPLICATION OF THE DEVELOPED MODELS IN STUDIES OF GEODYNAMIC PROCESSES 95
Kasza Damian

THREE-DIMENSIONAL MODEL OF THE RZECZKA OBJECT (THE RIESE COMPLEX, SOWIE RANGE) AS AN ENHANCEMENT OF THE COMPREHENSIVE GEOTOURISM POTENTIAL OF AN UNDERGROUND TOURIST ROUTE 102
Kadziolka Katarzyna, Kasza Damian, Milczarek Wojciech J.

Author Index

Statement of Peer review

In submitting conference proceedings to *Web of Conferences*, the editors of the proceedings certify to the Publisher that

1. They adhere to its **Policy on Publishing Integrity** in order to safeguard good scientific practice in publishing.
2. All articles have been subjected to peer review administered by the proceedings editors.
3. Reviews have been conducted by expert referees, who have been requested to provide unbiased and constructive comments aimed, whenever possible, at improving the work.
4. Proceedings editors have taken all reasonable steps to ensure the quality of the materials they publish and their decision to accept or reject a paper for publication has been based only on the merits of the work and the relevance to the journal.

Title, date and place of the conference

XXIIIrd Autumn School of Geodesy, September 21-22, 2017, Wałbrzych, Poland

Proceedings editor(s):

Jan B. Blachowski

Justyna Górniak-Zimroz

Damian Kasza

Date and editor's signature

21.06.2018

17, avenue du Hoggar - PA de Courtabœuf – BP 112 - 91944 Les Ulis Cedex A (France)
Tél. : 33 (0)1 69 18 75 75 - Fax : 33(0)1 69 07 45 17 – www.edpsciences.org

Application of observations of recent tectonic activity in the Świebodzice Depression (the Sudetes, SW Poland) in assessing seismic hazard in the Fore-Sudetic Monocline

Marek Kaczorowski[1,*], *Damian* Kasza[2], *Ryszard* Zdunek[1], and *Roman* Wronowski[1]

[1]Polish Academy of Sciences, Space Research Centre, 18A Bartycka St., 00-716 Warszawa, Poland
[2]Wrocław University of Science and Technology, Faculty of Geoengineering, Mining and Geology, 27 Wyb. Wyspiańskiego St., 50-370 Wrocław, Poland

Abstract. Tiltmeter observations with application of horizontal pendulums have been carried out for 40 years in the Geodynamic Laboratory in Książ. Long-term observations have not indicated any correlation of these data with meteorological or seasonal phenomena. Following an epoch of fast azimuth changes, a gradual compensation process took place, excluding the effect of gravitational creep of the rock massif. An assumption was made that the observed large changes of the equilibrium azimuths of the horizontal pendulums that result from tectonic tilt of the foundation blocks. Multiannual tiltmeter observations allowed to determine the temporal characteristics and amplitude tectonic effects. Intervals of strong tectonic activity in the rock massif of the Świebodzice Depression last from several days to over ten weeks and are separated by several tens of hours of low activity. Amplitude of the rock massif deformation reaches values from over ten to several tens of amplitudes of the tidal signal, i.e. up to several hundreds of micrometres. Water-tube tiltmeters (WT) launched in 2003 have confirmed the characteristics of tectonic effects and their incidental occurrence. Beside the tilt effects, WT have enabled to confirm vertical movement of the foundation blocks. Geological investigations in the Świebodzice Depression have indicated the presence of a numerous faults separating particular blocks in the rock massif. The presence of this fault system favours the dislocation of foundation blocks, which results in a quake-less relaxation of tectonic stresses and absence lack of seismic activity in the Świebodzice Depression. Foundation blocks separated by faults combined with the multiscale measurement system of WTs form a natural detector of regional tectonic activity, allowing to determine with micrometric resolution the representative function of tectonic activity in the rock massif of the Świebodzice Depression.

1 Location and natural conditions of the Geodynamic Laboratory at Książ

The Geodynamic Laboratory of the Space Research Centre of the Polish Academy of Sciences (GL) is located in the underground corridors under the Książ Castle [1-3]. The laboratory is situated at the depth of ca 50 m below the courtyard of Książ Castle and ca 40 m above Pełcznica river valley. The foundation of the castle hill is built of (Książ Formation) of Devonian and Carboniferous Age conglomerates [4] composed mainly of metamorphic gneisses from the nearby Sowie Mts [5-7]. Conglomerates of the Świebodzice Basin have been noted to occur in the large depths of the Earth's crust [6-8].

Connection of the Książ Massif with deep layers of the Earth's crust [7] results in good translocation of a wide spectrum of geodynamic signals: seismic, tidal, Earth's oscillations and tectonic signals. This is confirmed by continuous observations of strong, non-attenuated geodynamic signals registered by the GL instruments: horizontal pendulums and water-tube tiltmeters, relative and absolute gravimeters, vacuum interferometric extensometer, TM-71 crack gauge [9, 10], and wide-range seismograph of the Institute of Geophysics, Polish Academy of Sciences.

The GL underground excavations are characterized by a stable temperature (annual oscillations of ca. 0.1°C) and high relative humidity that varies seasonally at about 10% within 85% and 95%. The natural conditions of the laboratory provide a stable setting for the observation of subtle geodynamic signals.

2 Water-tube tiltmeters – the main source of geodynamic data

The first reports on the tectonic activity in the Świebodzice Depression (ŚD) were devoted to the foundation dip registered since the 1970-ties by quartz horizontal pendulums [11]. The determined temporal and amplitude characteristics have been confirmed by new instruments – water-tube tiltmeters (WT).

* Corresponding author: marekk@cbk.waw.pl

Due to the properties of horizontal pendulums such as instrumental drift, small dimensions of the measurement base and much lower sensibility than WT sensibility, pendulum tiltmeters are much less representative in the assessment of tectonic activity of the rock massif than the WT. At present, the source of information on the tectonic activity of the rock massif are two WT. For the determination of tectonic activity function two series of observations from WT were used (tectonic activity functions are defined in (see Chapter 3).

The main elements of WTs are two horizontal pipes that are perpendicular to one another, with a length of 65 and 92 m [12]. Interference gauges have been installed at the pipe ends, which register changes of water level related to the rock massif with regard to reference lenses with nanometric resolution.

The signal in form of a temporal chain of Newton rings is registered on hard discs and subsequently subjected to analysis that terminates with establishing a progressive function of the changes. The obtained measurement is relative to the rock massif, with which the reference lenses of the WT are firmly connected. A significant feature of WTs in measurements of the tectonic activity of a rock massif is their long measurement base. The WT measurement base (Fig. 1), i.e. the area from which the instruments collect information, is ca. 100 m x 100 m. Because this area is larger from the dimensions of single foundation blocks separated by faults, WT measurements supply information on the mutual vertical displacements and block tiltings, which guarantees high representativeness of the measurements in the assessment of momentous kinematic - tectonic activity of the rock massif.

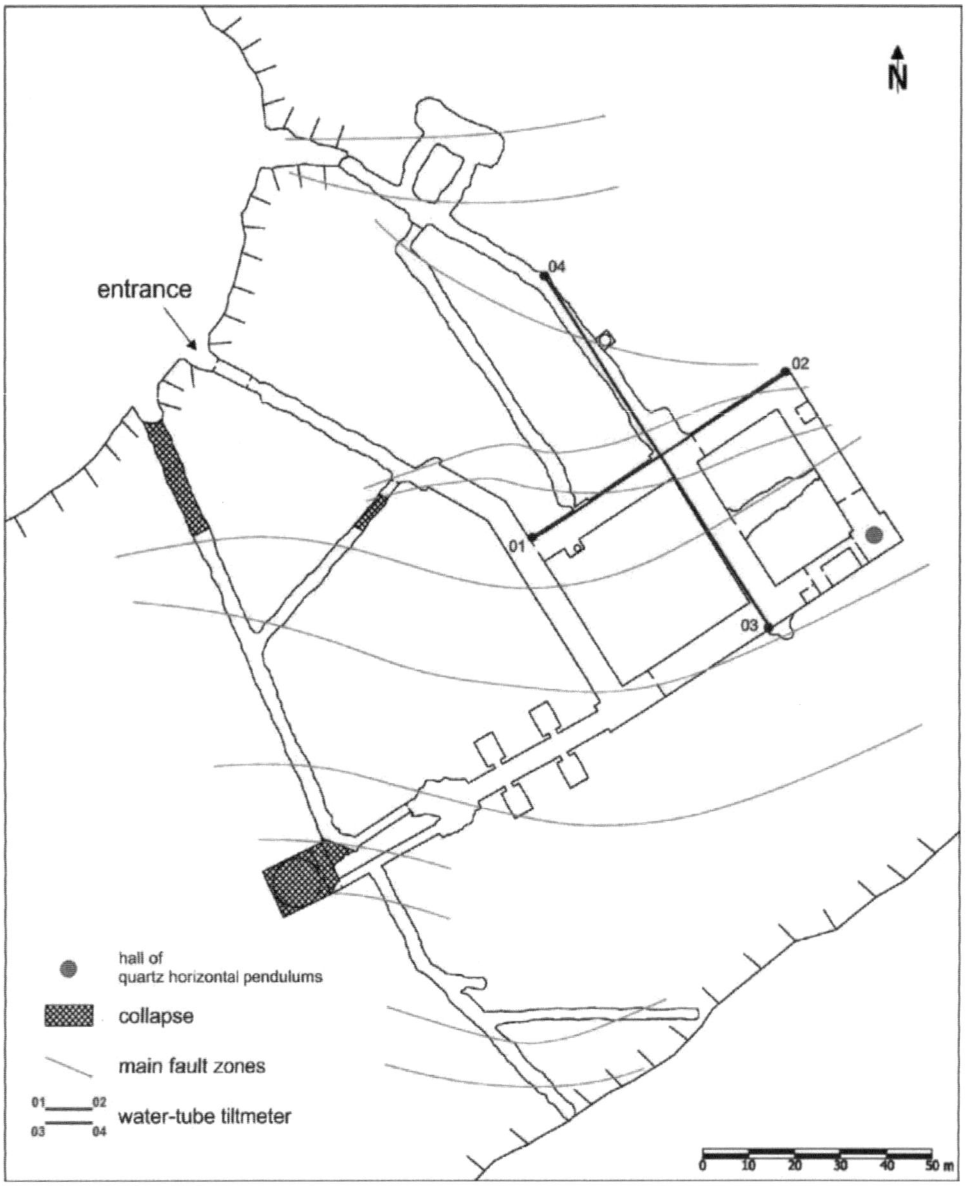

Fig. 1. Sketch-map of the underground excavations under the Książ Castle courtyard [13].

Changes of water level in the hydrodynamic systems of WTs are caused by two types of phenomena. The first type is related to external causes influencing Earth's gravity field, including loading effect, lunar and solar tidal phenomena as well as free oscillations of Earth's globe. The second type of phenomena registered by the WT are effects of tectonic origin [14-17]. The geodynamic activity of the rock massif is observed as tiltings effects and vertical displacements of the rock foundation [18]. Tectonic signals observed by WTs usually have much larger amplitude than the tidal signals. In extreme cases the tectonic signal exceeded the tidal signal several hundred-fold (in 2009 and 2010 on channel 03).

3 Definition of the function of tectonic activity of the rock massif (TAF) and its determination

The function of tectonic activity (TAF) was introduced to describe the magnitude and temporal changes of tectonic activity of the rock massif. This is an empirical function determined on the basis of measurements by WTs or other instruments with a sufficiently high sensibility for the deformations of the rock massif and a wide measuring base [19]. The tectonic activity registered by WTs is represented by two types of deformation: the tilt of the entire rock massif (argument a) and vertical displacements of the foundation blocks (function $r(l)$), where a is the tilt of the massif and $r(l)$ is a function of the radial deformation of the foundation rock massif under the pipe of the tiltmeter [20]. The signals related to the vertical displacement of the rock massif are the integral of the vertical displacements of the foundation under WT pipes. WT measuring systems form a non-linear superposition of both types of deformation, whose mutual proportions vary in time. The tectonic signal registered by WTs is presented by the following formula:

$$S(a, r(l)) = L * \cos(a) + \int_{l=0}^{l=L} r(l)dl \qquad (1)$$

where L is the length of the tiltmeter pipe.

The TAF function is obtained from the $S(a,\ r(l))$ formula in several steps. In the first step, jumps and discontinuities caused by instrumental reasons or intense seismic phenomena are removed from the raw observation set. After removal of jumps and discontinuities, the effects of water evaporation and high-frequency signals are eliminated. Next, the measurement series are subject to tidal adjustment with help of ETERNA 3.4 software in order to remove the main part of the tidal signals and to determine the tectonic signal [21].

After additional adjusting by polynomial splines, TAF is obtained for each measurement channel of the WTs. Additional adjusting of the observation data is required due to the application of TAF for the determination of the function of deformation rate of the rock massif, i.e. VTAF - derivative of TAF.

4 Temporal and amplitude relationships between tectonic activity of the Świebodzice Depression and the seismic activity of the Fore-Sudetic Monocline

Elaboration of the observation series from WTs and determination of TAF derivatives allowed to discover the temporal and amplitude relationships between the rock massif activity of the Świebodzice Depression and the strong (>=4 Mag) seismic events in the Fore-Sudetic Monocline [22]. Comparative analysis of both types of phenomena lead to a conclusion that strong seismic shocks take place only at specific stages of tectonic activity of the ŚD rock massif. Stages of tectonic activity of the rock massif are described by the TAF functions and their derivatives (Fig. 2). Two basic intervals may be distinguished in the observed deformation process of the ŚD rock massif: interval of high rates of deformations and interval of low rates of deformations, during which the rates of deformations are close to zero, to later change their turn.

In the interval when the deformation stops and its rate changes sense, i.e. the rock massif transforms from the first to the second phase, the rock massif is in the state of extension - stretching for a short while. During the extension, lasting for a few to several tens of hours, significant changes take place in the rock massif, such as thickness increase of the dislocation zones, opening of fractures and fissures, gases outflow and decrease of the mean density of the rock massif. During the compression, the rock massif is subjected to high deformation rates lasting from several days to several tens of days, which are separated by extension phases lasting several tens of hours (Fig. 2).

The processes which take place in the rock massif during extension intervals, particularly the decrease of the mean density of the rock massif (dilution of the rock massif) favour the possibility of earthquake occurrence. The earthquakes are accompanied by rock massif destruction in form of fractures, crushing, strain, etc., causing volume increase and decrease of the mean values of rock massif density.

Based on these assumptions, a seismic shock occurs under the condition of existence of additional space within the rock massif, indispensable for the absorption of crushed material produced during the earthquake. This missing space is created when the rock massif is in the extension phase. Therefore, during the extension phase the rock massif becomes susceptible to destruction.

Comparative analysis of multiannual sequences of the TAF function with temporal distribution of strong seismic events confirms the above reasoning. All the strong seismic phenomena (>= 4 Mag, 26 seismic events from the period of 2015-2016) took place at low rates of rock massif deformation, i.e. during extension phases.

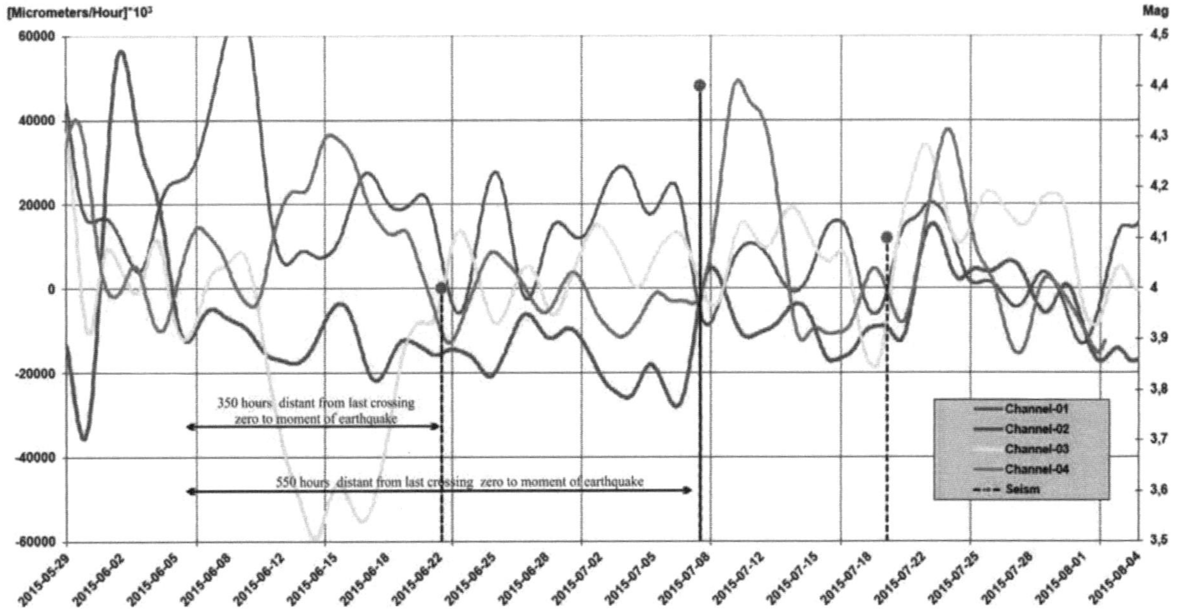

Fig. 2. 2015 – plots of derivatives of tectonic activity functions and the three strongest shocks: 4.0, 4.1 and 4.4 Mag.

4.1 Temporal and amplitude rules of the process at which the rock massif is ready for a seismic shock

The results of comparative analysis made for 26 seismic events (from 2015-2016; EMSC, 2017) allowed to conclude that the processes in the rock massif which precede seismic events follow a repeatable succession of rules. This observation allowed to include in the description of the process at which the rock massif becomes ready for a seismic shock, the temporal and amplitude rules consisting of a number of precedents.

Temporal precedents determine the time presented in hours since the last change of the sign of the TAF derivative on each WT measurement channels to the seismic shock. For every case of strong seismic event (>= 4 Mag) from the years 2015-2016, the temporal characteristics correspond to the precedents described below (i.e. Fig. 3):

– **Precedent T-1**: 4 passages through zero of the derivatives on all channels within ca. 100 h before the shock (very strong condition –very highly probable event during the next 100 h)

– **Precedent T-2**: 3 passages through zero of the derivatives within ca. 100 h before the shock and preceding the passage up to 600 h before the shock (strong condition – seismic shock highly probable during the next 100 h)

– **Precedent T-3**: 2 passages through zero within ca. 100 h before the shock and 2 preceding passes before 400 h before the shock (medium-strong condition – shock probable during the next 100 h)

– **Precedent T-4**: 2 passages through zero within ca. 100 h before the shock and 2 symmetric channels postponed to < 600 h.

Precedent T-5: 1 passage through zero within ca. 100 h before the shock and 3 focused passages up to 400 hours before the shock (medium-strong condition – shock probable during the next 100 h).

The temporal intervals applied in the time precedents have been determined empirically.

Amplitude characteristics contain information of the values of the TAF derivatives at the moment of the seismic event observed on the 4 WT measurement channels. For all strong seismic events the amplitudes characteristics are described by the following amplitude precedents: four-fold concentration, three-fold concentration or two-fold concentration of the TAF derivative values in the interval of –20000 to 20000 [μm/h] (i.e. Fig. 4; their magnitudes result from the 103 [μm/h] multiplier used to operate on integer numbers):

– **Precedent A-1**: four-fold concentration close to zero (values of derivatives from –22000 to 22000 [μm/h] – very strong condition – high probability of seismic shock within 100 h.

– **Precedent A-2**: three-fold concentration close to zero (values of derivatives from –22000 to 22000 [μm/h] and 1 channel, in which value of derivative is <60000 [μm/h] – strong condition – high probability of seismic shock within 100 h.

– **Precedent A-3**: two-fold concentration close to zero and two-fold close concentration with values of derivatives from –25000 to 25000 [μm/h] – strong condition – high probability of seismic shock.

– **Precedent A-4**: two-fold concentration close to zero with values of derivatives from –22000 to 22000 [μm/h] and two-fold concentration for values <60000 [μm/h].

– **Precedent A-5**: two-fold concentration close to zero with values of derivatives from –20000 to 20000 [μm/h] and two channels with mutually symmetrical values at a distance of <80000 [μm/h].

– **Precedent A-6**: double two-fold concentrations with values of derivatives from –22000 to 22000 [μm/h] symmetrical one to the other; high probability of seismic shock.

The numerical intervals applied in the definitions of amplitude precedents have been determined empirically.

Comparative analysis indicates that the essential condition for the occurrence of a seismic shock is the realization of temporal and amplitude precedents (Figs. 3, 4). However, precedent fulfillment is not a condition sufficient for the occurrence of a seismic shock. In some cases, despite precedent fulfillment, a seismic shock did not occur in the extension phase.

We observed that before a seismic shock the TAF derivative change sign on all 4 WT measurement channels in an interval not exceeding 480 h before the seismic event. The inflection points of the TAF derivatives and an earlier change of the sign of the derivatives (>780 h), and the case when the derivative function is tangent to zero usually do not result in a shock. In the period 2015-2016 all 26 seismic shocks (>=3.6 Mag) were preceded by the fulfillment of temporal and amplitude precedents by the TAF derivatives (see Fig. 3, 4, 5, 6 and Tables 1 and 2).

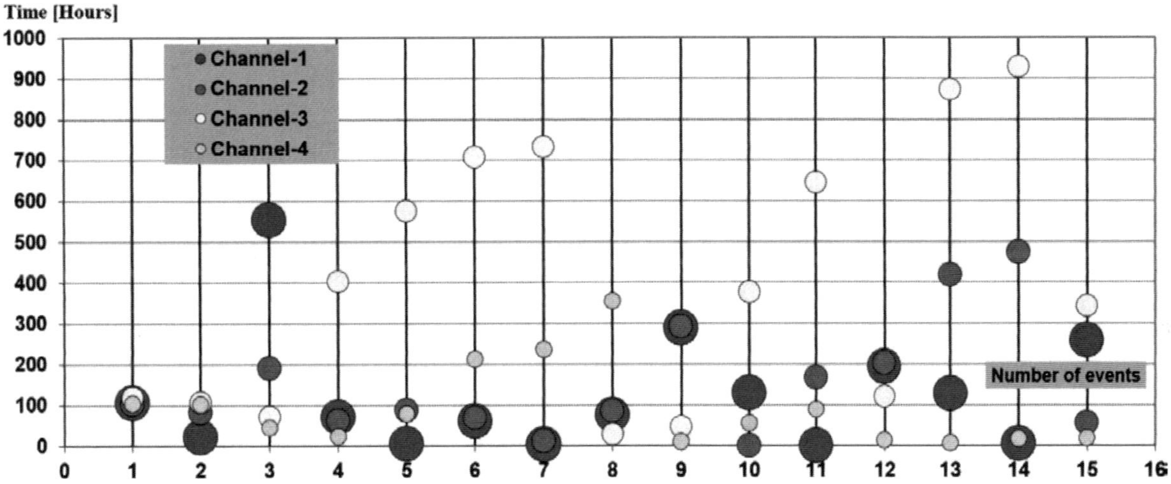

Fig. 3. 2016 – conditions for shock occurrence for 15 seismic events; temporal precedents T.

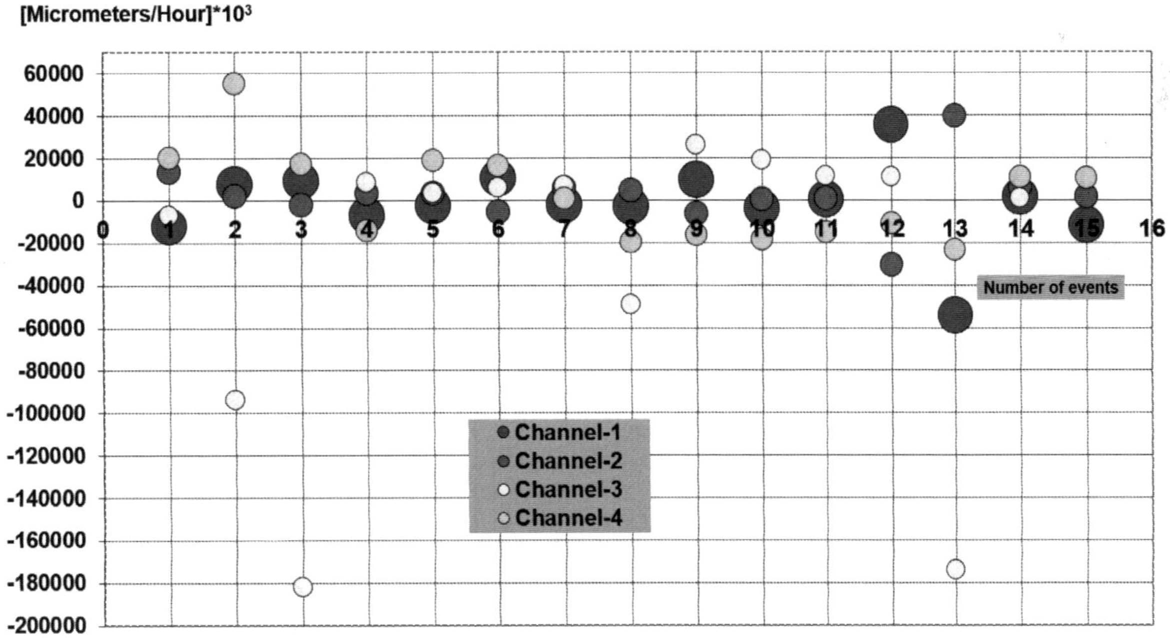

Fig. 4. 2016 – conditions for shock occurrence for 15 seismic events; amplitude precedents A.

5 Comparison of TAF derivatives with strong seismic events in the Fore-Sudetic Monocline in 2015-2016

In 2015, eleven strong seismic shocks with magnitudes from 3.6 to 4.4 Mag took place. All shocks occurred during the extension phase. No strong seismic shocks took place apart from the extension intervals (Tab. 1, Figs 2, 5, 6).

In 2016 there were 15 strong shocks with magnitudes from 3.6 to 4.4. All shocks occurred during the extension phases. No strong seismic shocks took place apart from the extension intervals (Tab. 2, Figs. 3, 4, 7).

Table 1. Temporal and amplitude precedents for 11 seismic events with magnitudes from 3.6 to 4.4 in 2015 [22].

No.	Date	Mag	Amplitude precedents A	Description of amplitude precedents	Temporal precedents T	Description of temporal precedents
1	2015-02-05	3.9	A-1B15	Four very close to zero, two-fold symmetry medium + concentration	T-1A	Four tight – close to seismic moment
2	2015-02-12	3.8	A-1B15	Four very close to zero, ASYMMETRY + three-fold concentration	T-1A	Four tight – close to seismic moment
3	2015-06-22	4	A-1B15	Four very close to zero, ASYMMETRY + three-fold concentration	T-4A	Single groupings Type I
4	2015-07-08	4.4	A-1B15	Four very close to zero, ASYMMETRY + three-fold concentration	T-3A	Three close to zero tight + one distant
5	2015-07-19	4.1	A-1B15	Four very close to zero, ASYMMETRY + three-fold concentration	T-3A	Three close to zero tight + one distant
6	2015-09-09	4	A-2B26	Three very close to zero, ASYMMETRY + three fold - strong concentration – grouping nos. 212, 262	T-4A	Single groupings Type I
7	2015-10-29	4	A-2A20	Three close to zero, very strong symmetry (0.8 and 0.8)	T-4A	Single groupings Type I
8	2015-11-20	3.7	A-2A21	Three close to zero, ASYMMETRY + three-fold slight concentration – grouping nos. 211, 261	T-3A	Three close to zero tight + one distant
9	2015-12-11	3.8	A-1B15	Four very close to zero, ASYMMETRY + three-fold concentration	T-4B	Two close to zero tight + two tight
10	2015-12-28	3.7	A-1B15	Four very close to zero, ASYMMETRY + three-fold concentration	T-3A	Two close to zero tight
11	2016-01-16	3.7	A-1B15	Four very close to zero, ASYMMETRY + three-fold concentration	T-4A	Three close to zero tight + one distant

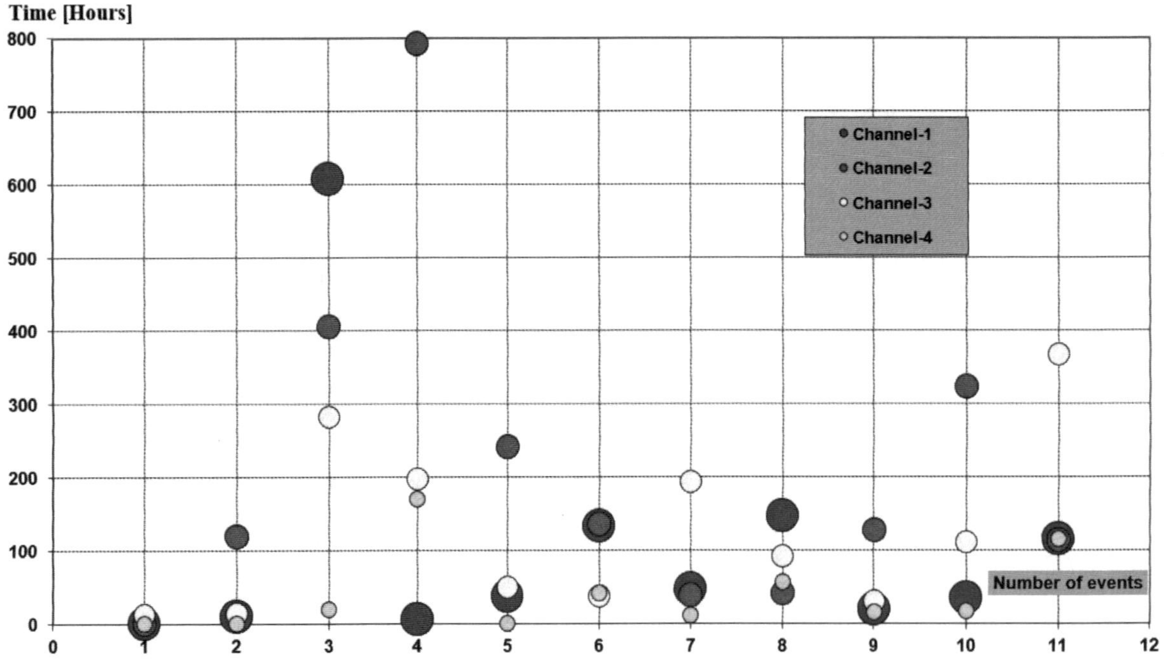

Fig. 5. The year 2015 – conditions for shock occurrence for 11 seismic events. Temporal precedents T.

 https://doi.org/10.1051/e3sconf/20185500001

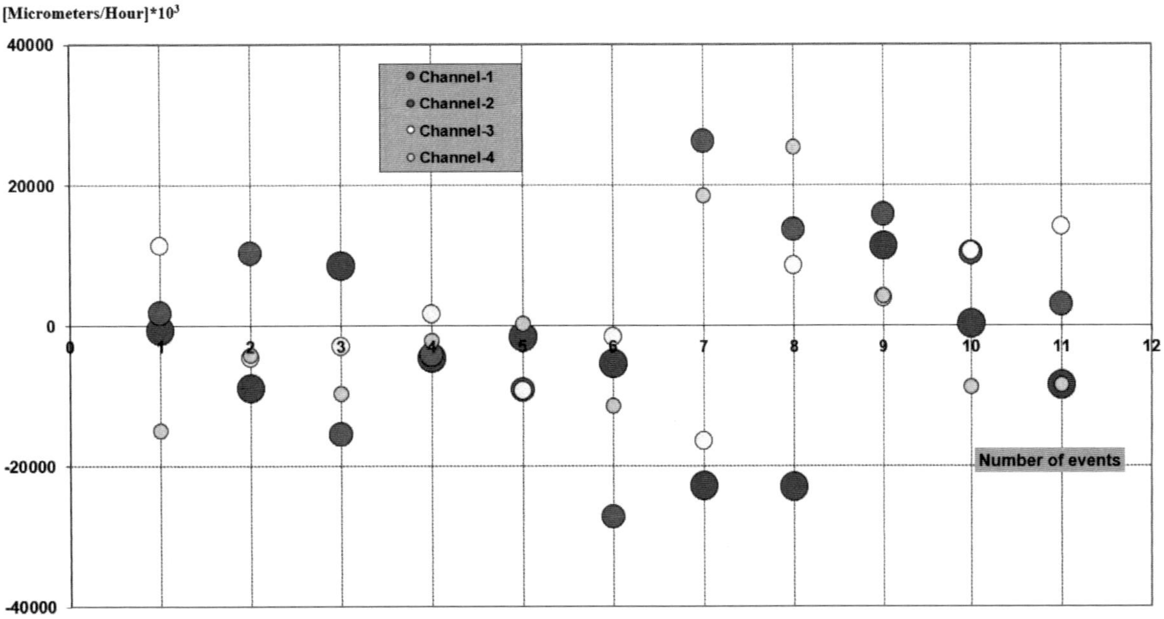

Fig. 6. The year 2015 – conditions for shock occurrence for 11 seismic events. Amplitude precedents A.

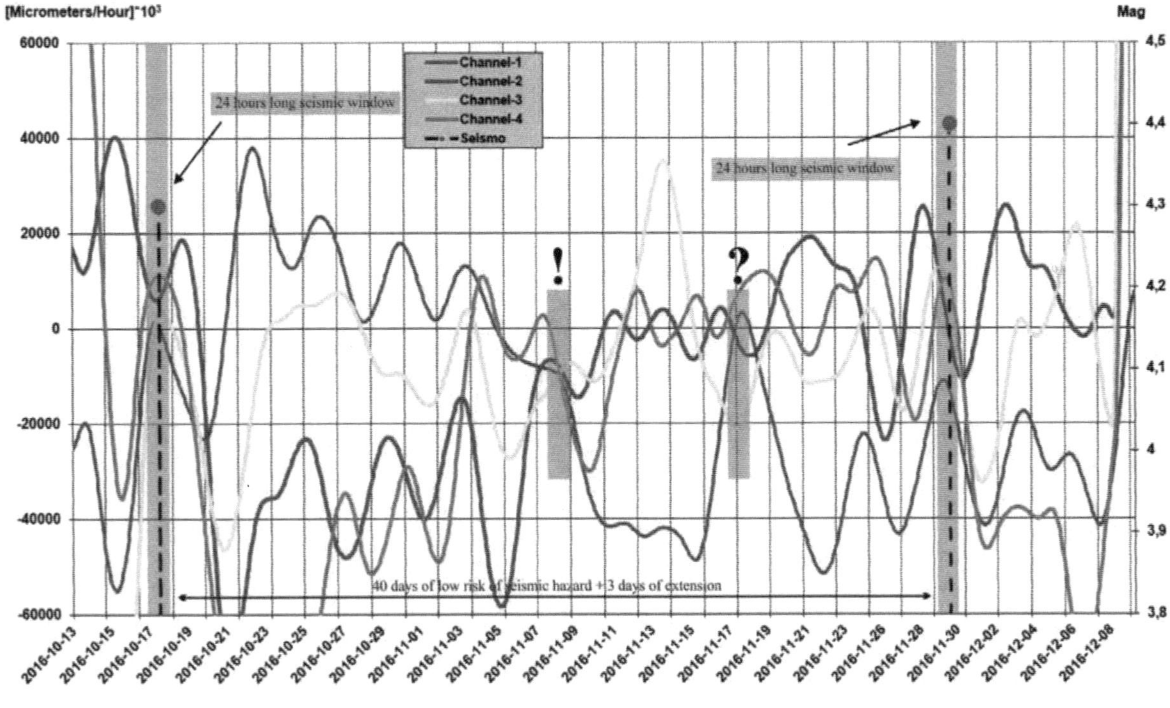

Fig. 7. 2016 – Plots of derivatives of tectonic activity functions and two strong shocks in: 2016-10-17 4.3 Mag and 2016-11-29 4.4 Mag.

 https://doi.org/10.1051/e3sconf/20185500001

Table 2. Temporal and amplitude precedents for 15 seismic phenomena with magnitudes from 3.6 to 4.4 in 2016 [22].

No.	Date	Mag	Amplitude precedents A	Description of amplitude precedents	Temporal precedents T	Description of temporal precedents
1	2016-02-25	4.1	A-1B15	Four very close to zero, ASYMMETRY + two concentrations	T-1A	Four tight – close to seismic moment
2	2016-03-08	3.6	A-3B36	Two very close to zero, strong single symmetry – groupings nos. 308, 358	T-1A	Four tight – close to seismic moment
3	2016-04-08	3.7	A-2B25	Three very close to zero, very poor symmetry	T-4A	Single grouping Type I
4	2016-04-28	3.6	A-1B15	Four very close to zero, medium symmetry + concentration	T-3A	Three close to zero tight + one distant
5	2016-05-05	3.7	A-1B15	Four very close to zero, ASYMMETRY + three-fold concentration	T-3A	Three close to zero tight + one distant
6	2016-05-11	3.9	A-1B15	Four very close to zero, ASYMMETRY + three-fold concentration	T-4A	Single grouping Type I
7	2016-05-12	3.8	A-1B15	Four very close to zero, ASYMMETRY + three-fold concentration	T-4A	Single grouping Type I
8	2016-06-02	4.1	A-2B26	Three very close to zero, ASYMMETRY + poor three-fold concentration – groupings nos. 211, 261	T-3A	Three close to zero tight + one distant
9	2016-07-30	4.4	A-2B25	Three very close to zero, poor symmetry (0.5 and 0.5)	T-4B	Two close to zero tight + two tight
10	2016-08-13	4.3	A-1B15	Four very close to zero, very poor symmetry	T-3A	Two close to zero tight
11	2016-08-24	3.8	A-1B16	Four very close to zero, ASYMMETRY + single-fold concentration	T-4A	Three close to zero tight + one distant
12	2016-09-14	4.1	A-3B35	Two very close to zero, very strong two-fold symmetry– groupings nos. 304, 354	T-4B	Single grouping Type I
13	2016-10-15	3.7	A-4A30	One close to zero –DETAILS missing– groupings nos. 30, 35	T-4A	Two close to zero tight + two distant
14	2016-10-17	4.3	A-1B15	Four very close to zero, ASYMMETRY + three-fold concentration	T-4A	Two close to zero tight
15	2016-11-29	4.4	A-1B15	Four very close to zero, ASYMMETRY + three-fold concentration	T-4B	Two close to zero tight + two tight

5 Summary and conclusions

Comparative analysis of the tectonic activity of the Świebodzice Depression rock massif with the seismic events in the Fore-Sudetic Monocline indicated that all 26 strong seismic shocks (>= 3.6 Mag) occurred in the years 2015-2016 during the extension phases, i.e. at the low rates of the rock massif deformations. In the compression phases, i.e. at high rates of rock massif deformation, no strong seismic shock (>= 3.6 Mag) happened. Detection of the high temporal concordance of the massif extension intervals in the Świebodzice Depression with the 26 strong seismic events (>= 3.6 Mag) in 2015-2016 provides significant arguments to support the concept of a large-scale, homogeneous field of tectonic forces, covering two geological units, i.e. the Świebodzice Depression and the Fore-Sudetic Monocline at the same time.

Comparative investigations of the tectonic activity of the Świebodzice Depression with the seismic events in the Fore-Sudetic Monocline lead to the conclusion that the processes in the rock massif, which precede seismic event, follow a repeatable succession of rules.

Gathered observations allowed to introduce the description of the process of adaptation of the rock massif to fulfill the conditions for seismic shock occurrence. This description comprises temporal and amplitude rules, composed with a number of precedents.

We noticed the essential condition for the occurrence of an earthquake in the Monocline which is the rule of passage through zero of four derivatives of the TAF at ca. 100 h before the shock and the rule of concentration around zero of values of most derivatives on the most channels within the range 22000 and –22000 (μm/h*1000; Tables 1 and 2).

Results of comparative analysis indicate that after fulfil temporal and magnitudes rules seismic hazard in the Fore-Sudetic Monocline increase. The presented method allowed for determining the intervals of low rates of the rock massif deformation, i.e. the intervals of extension within the years 2015-2016, during which 26 seismic events (>=3.6 Mag) took place. A stronger shock results in better application of the determined rules.

A regularity was noted that the correlation of low-energy shocks with intervals of low rates of deformations (extension intervals) is worse than correlation of high-energy shocks. This observation suggests that probable the low-energy shocks are generated by mining activity in contrast to the high-energy shocks of tectonic nature which are well correlated with tectonic activity of ŚD.

For unknown reasons, in a few extension intervals strong seismic shocks did not take place despite the fulfilment of temporal and magnitude precedents conditioning the occurrence of earthquakes. The acquired knowledge allows for recognising almost all strong seismic events (>= 3.6 Mag), which enables us to forecasting of the seismic hazard in the Fore-Sudetic Monocline. It is worth mentioning that the presented idea is not applicable for unambiguous prediction of the earthquakes.

This work was financed by the Polish Statutory Research Grant no. 0402/0137/16 and MNiSW no. 6158/E-73/SPUB/2016/3-1.

References

1. B. Rdułtowski, *The Hunt for Wunderwaffe – Hitler's Secret Weapon from the Owl Mountains* (Technol, 2012)
2. "History of AL Riese", Gross-Rosen Museum in Rogoźnica. http://en.gross-rosen.eu/historia-kl-gross-rosen/historia-al-riese/, retrieved 7 October 2017
3. P. Zagożdżon, K. Zagożdżon, Hereditas Minariorum **3**, 267-279 (2016)
4. H. Teisseyre and O. Gawroński, Detailed Geological Map of the Sudetes, 1:25,000, Świebodzice sheet, Państwowy Instytut Geologiczny, Warszawa (1965)
5. W. Nemec, S.J. Porębski, R.J. Steel, Sedimentology, 27, 519-538 (1980)
6. S.J. Porębski, Geologia Sudetica, **16**, 1, 99-190 (1981)
7. S.J. Porębski, N. Jb. Geol. Paläont. Abh., **179**, 259-274 (1990)
8. A. Żelaźniewicz, P. Aleksandrowski, Przegląd Geologiczny, **56**, 10, 904-911(2008; in Polish with English abstract)
9. D. Kasza, M. Kaczorowski, R. Zdunek, R. Wronowski, 2014, Acta Geodynamica et Geomaterialia, **11**, No. 3 (175), 225-234 (2014)
10. R. Zdunek, M. Kaczorowski, D. Kasza, R. Wronowski, Acta Geodynamica et Geomaterialia, **11**, No. 4 (176), 305-315 (2014)
11. T. Chojnicki, P.A. Blum, Artificial Satellites, vol. **31**, no. 3, 123-129 (1996)
12. M. Kaczorowski, *Earthquake Source Asymmetry, Structural Media and Rotation Effects* (Springer-Verlag, 493-520, 2006)
13. D. Kasza, PhD dissertation, Wrocław University of Science and Technology, unpublished (2016; in Polish)
14. M. Kaczorowski, Acta Geodynamica et Geomaterialia, **4**, 4, 109-119 (2007)
15. M. Kaczorowski, Reports on Geodesy, **85**, 2, 79-86 (2008)
16. M. Kaczorowski, Acta Geodynamica et Geomaterialia, **6**, 3, 369-381 (2009)
17. M. Kaczorowski, Bulletin d'Information des Marées Terrestres (BIM), **144**, 11605-11613 (2009)
18. M. Kaczorowski, J. Wojewoda, Acta Geodynamica et Geomaterialia, **8**, No. 3(163), 1-13 (2011)
19. H.E. Ruotsalainen, Lithosphere 2008 symposium Oulu, Finland, extended abstracts, Report S-53, Institute of Seismology, University of Helsinki (2008)
20. D.R. Bower, Phil. Trans. R. Soc. Lond. A, **274**, 223-226 (1973)
21. H.-G. Wenzel, Obs. Royal de Belgique, Bull. Inform. Marées Terr., **124**, 9425–9439 (1996)
22. European Mediterranean Seismological Centre, https://www.emsc-csem.org, retrieved 7 October 2017

Enhancement of the Walbrzych Hard Coal Mines Geographic Information System for its application in studies of mining deformations

Jan Blachowski[1,*], and *Paulina* Herkt[1]

[1]Wrocław University of Science and Technology, Faculty of Geoengineering, Mining and Geology, 27 Wyb. Wyspiańskiego St., 50-370 Wrocław, Poland

Abstract. Geographic information systems (GIS) have been increasingly used in augmenting studies of mining and post-mining areas such as surface deformation analyses, risk assessment associated with old mining sites, identification of old mining sites and other. Up-to-date and complete geodatabase is an indispensable element of any geoinformation system facilitating such geospatial analyses. In this paper the process of verification and integration of Walbrzych Hard Coal Mines geodatabase and project of enhancing the Deformation Information System for these mines with web map application of underground workings and query tools developed with free and open source software (FOSS) have been described. The interactive map allows users to interact and obtain precise information on location and characteristics of underground mining and the query tools streamline pre-processing operations necessary for geospatial analyses.

1 Introduction

Geographic information systems (GIS) have been increasingly used in augmenting studies of mining and post-mining areas. The flexibility of GIS in handling and processing of large datasets, varied in terms of sources and formats, as well as GIS analytical capabilities have been used to solve problems associated for example with processing of satellite radar interferometry (InSAR) results for analysis of ground movements in areas of former underground hard coal mining [1-4], risk assessment associated with destruction of old underground workings and subsequent appearance of discontinuous deformations on surface [5-8], identification of abandoned underground mines [9, 10] and other issues.

In case of old and now abandoned underground mines development of GIS database with complete geometrical and attribute information on former mining activity in a given area requires additional work and attention. This is because, in many cases data sources, usually old paper mine plans or maps may be lost, incomplete or damaged and the digital coding of such documents is time consuming. Depending on the data available, needs and GIS skills of researchers, geoinformation systems used in studies of former mining sites differ in functionality, ranging from simple digital data repositories and visualization tasks to advanced systems with tools and functions for complicated analyses.

This paper presents the current status of development of the Deformation Information System for the underground hard coal mines in Walbrzych with special focus on development of interactive web map with free and open source software (FOSS) and query tools to facilitate preprocessing operations associated with surface deformation analyses.

2 Geology and coal mining in the Walbrzych Coal Basin

It is estimated that hard coal mining in Wałbrzych started in the 14th Century. At that time coal was mined from the surface. With the progress of time underground methods were introduced and coal production intensified. In the second half of the 19th Century mining of mineral from coal fields located under human settlements and infrastructure (roads, railways, etc.) began. After the World War II coal production peaked in 1955 with 3,25M tonnes, later on the values oscillated around 2,5M tonnes per year. Mining gradually ceased between 1993 and 1999. The steady drop in production of coal was accompanied by controlled flooding of underground workings through decrease in mining drainage of rock mass. The process of mine flooding was started in 1994 and because of favourable geological settings particular coal fields could have been flooded independently [11].

The Wałbrzych coal basin is of a limnic-type, that is deposition of material took place in intra-mountainous sedimentary basins. This has resulted in a mixed exogenous – endogenous coal deposit [12]. The geology of the basin is complicated in terms of deposit

* Corresponding author: jan.blachowski@pwr.edu.pl

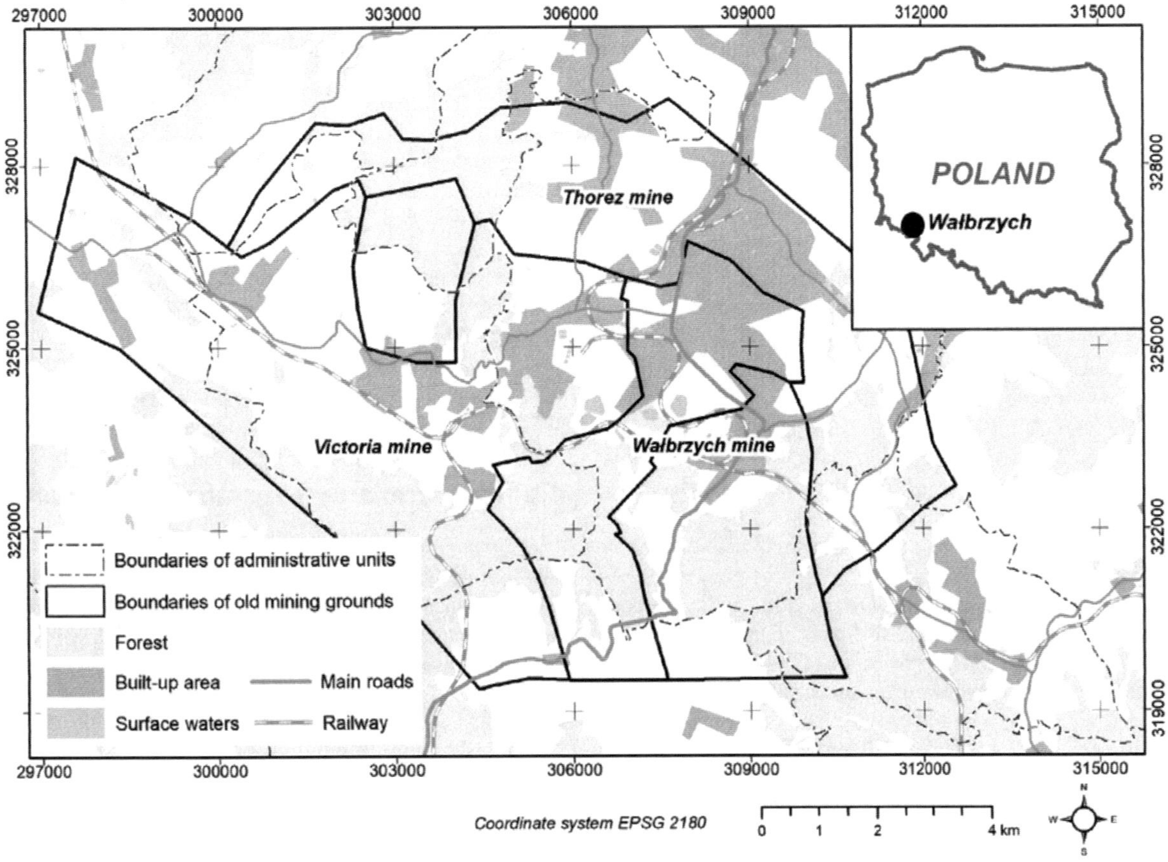

Fig. 1. Location of the old Hard Coal Mines in Walbrzych.

continuity, inclination, thickness and tectonic conditions [13]. The coal layers are associated with three of the four lithostratigraphic Pennsylvanian complexes, i.e.: Žacléř Biały Kamień and Wałbrzych formations. Altogether, 80 coal seams have been documented, including 48 in the Žacléř and 30 in the Wałbrzych formations. The coal layers are inclined in the direction of the centre of the basin, and the dip ranges from several to over 30 degrees, and even 60 degrees in the basin edges. Most of numerous faults trend from NW towards SE with the influence of the Chelmiec intrusion clearly marked by longitudinal and latitudinal faults. The throw of the main faults reaches 300 m. In addition, there are numerous local faults in the coal-bearing layers, with throws of several metres [12].

Difficult mining conditions were the main reason for high costs of mining and closure of mines in Walbrzych. Location of the old coal basin has been shown in Fig. 1.

3 Chronology of development of geoinformation system for the Walbrzych Hard Coal Mines

This work is a continuation of studies started in the Wroclaw University of Science and Technology in 2006 and presented in papers by Blachowski [14], Blachowski and Stefaniak [15], Blachowski and Stefaniak [16] and Blachowski et al. [17]. The first one focused on development of a GIS database structure for storage and handling of numerical data related to underground mining of hard coal in the Walbrzych area. The results include a logical database structure and geodata describing coal mining for two of the three Walbrzych mines, i.e. the *Thorez/Julia* and the *Victoria* mines [14]. The succeeding work [15] produced geodata characterizing coal mining in the third mine *Walbrzych* and the remaining coal field *Barbara* in the *Victoria* mine. The study was also aimed at development of GIS procedures augmenting studies of ground deformation in the city of Walbrzych such as search and query tools according to selected criteria. The criteria being: depth of mining, mining method, coal panel thickness, coal panel inclination, and time of mining. This work was followed by a study on the application of self-organising maps (SOMs) to explore and analyse geodata describing factors associated with ground subsidence and study influence of particular ground subsidence factors on the observed surface subsidence, as well as the relationships between these factors [16]. It has been found that exploratory spatial data analysis is useful to identify relationships in multidimensional data related to mining induced ground subsidence.

These studies have resulted in a concept of a GIS based deformation information system for studies of ground movements on mining and post-mining grounds. The concept has been presented in a paper published in 2014 [17]. The proposed system's architecture developed for the Walbrzych mines consists of data collection, data visualization components, as well as spatial data mining, modelling and classification modules. The core of the system is a geodatabase storing vector and raster format data related to mining, geology, deformation surveying, landuse, etc. The analytical tools developed and implemented in the system include functions:

- to interpolate subsidence surfaces from discrete point measurements, assess quality and accuracy of interpolations and store them in raster format,
- to calculate basic parameters characterising ground deformation in space and in time such as: tilt, curvature, and horizontal strain,
- to classify area based on these parameters and in accordance with mining ground classification used in Poland, spatial regression functions to model and predict ground subsidence, and
- functions for data preprocessing for external applications.

The main purpose of the system is to facilitate and support studies of ground deformation during and after end of coal mining in the city of Walbrzych. The system's application have been presented in papers by Blachowski and Milczarek [18], and by Blachowski [19], as well as in doctoral dissertation by Milczarek [20]. The first one analysed surface changes in the Walbrzych mining grounds that occurred between 1886 and 2009. The analysis has been based on comparison of historical and present-day topographic maps and map algebra functions in GIS. The second, utilized spatial regression methods to determine relationship between subsidence (dependent variable) and subsidence factors (independent variables). The model has been then used o predict subsidence in areas that have not been surveyed. The last one used the system's geodatabase for construction of models for deformation calculations with finite element method (FEM).

In an associated study by Blachowski and Nowacka [21], an interactive web map application for presentation of information on remaining and now demolished buildings and shaft towers of one of the former mines - Thorez has been developed.

In the following chapters, methodology and results of current work on database integration and web presentation of underground workings has been described. Verification of database completeness and integrity has been done within M.Sc. diploma thesis [22] at the Wroclaw University of Science and Technology.

4 Verification and integration of geodata

The main source of information on coal mining in the past comprised of mining maps and associated geological and mining documentations. Each coal layer has been mapped at various scales: 1:1 000, 1:2 000 and 1:5 000 and each layer has been represented by sections. The number of sections ranged from few to several dozen for each coal layer depending on map scale. Overall, 29 coal levels have been digitised for the *Victoria* coal mine, 28 coal levels for the *Walbrzych* coal mine and 18 for the *Julia/Thorez* coal mine. In total several hundred map sections have been examined, georeferenced and digitised to a vector format with additional coding of attributes that describes mining. Fragment of an example map section with geometry and attribute information that could be interpreted has been shown in Fig. 2.

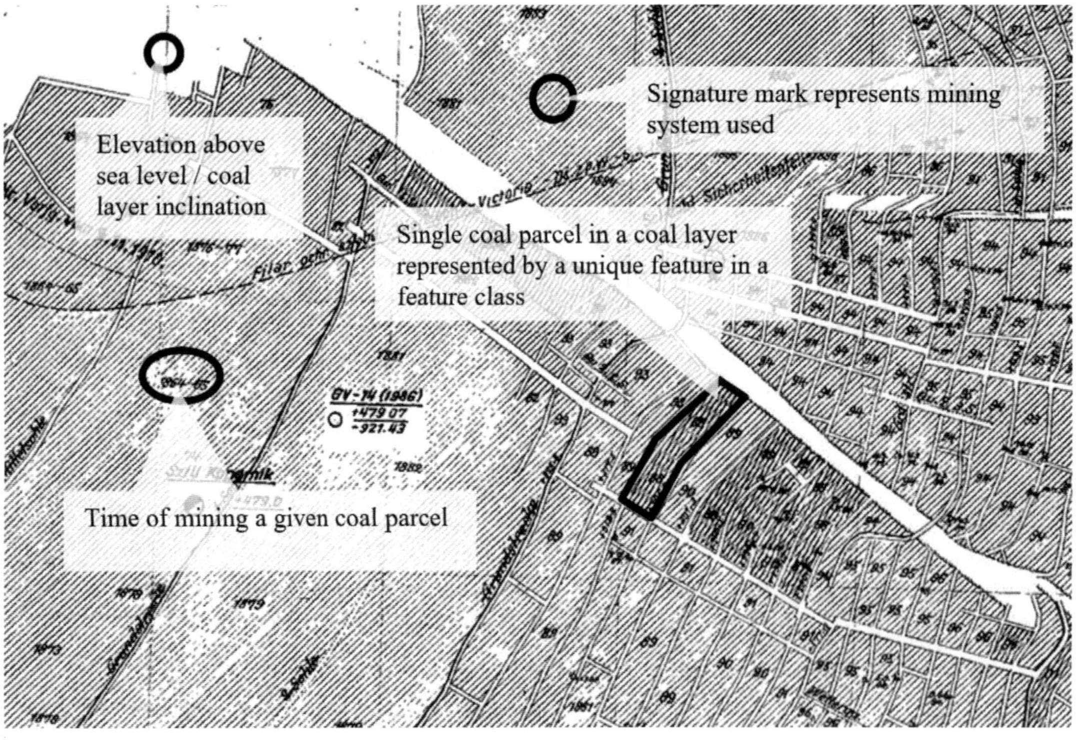

Fig. 2. Sample section of mine plan used for geodatabase development.

https://doi.org/10.1051/e3sconf/20185500002

Within the presented project the GIS data developed for each mine have been integrated in a common geodatabase. In this process field names of feature classes have been standardized, geometry and topology of features has been checked, as well as attributes checked for missing or wrong values. The final database structure comprises of polygon feature classes representing each a given coal layer with the following attribute information associated with each feature class object (coal parcel):

- Start time (of mining), date format,
- End time (of mining), date format,
- Mining system, string format,
- Mining system (code), string format,
- Coal layer thickness, number format,
- Coal parcel area, number format,
- Coal parcel volume, number format.

Other feature classes stored in geodatabase represent: location of mine shafts, location of main galleries and drifts, boundaries of mine shaft protection areas, boundaries of old mining grounds, as well as geodetic data representing surface deformations.

The next stage of the project focused on developing query tools to find and select coal parcels that meet given criteria. The search and query tools have been developed with Python language and implemented as a Python Toolbox in ArcGIS platform. Three query algorithms have been prepared, (1) query based on single parameter, (2) query based on a single parameter and a given range of values, and (3) query based on three parameters.

The first tool is a simple query returning features that meet a value provided, e.g. all parcels that were mined with a selected mining system.

In the case of (2) user can query the following criteria: time of mining (start and end time), thickness (minimum and maximum), as well as area and volume (minimum and maximum). In addition, two values that define the range are required, e.g. parcels mined from 1900 to 1910.

The third tool is the most complex and allows to select features that meet three criteria, including two based on a range of values, e.g. features representing coal parcels mined with wall system with dry fill, thickness from 1.0 m to 1.5 m and extracted between 1900 and 1950.

Graphical user interface of tool with three parameters has been presented in Fig. 3.

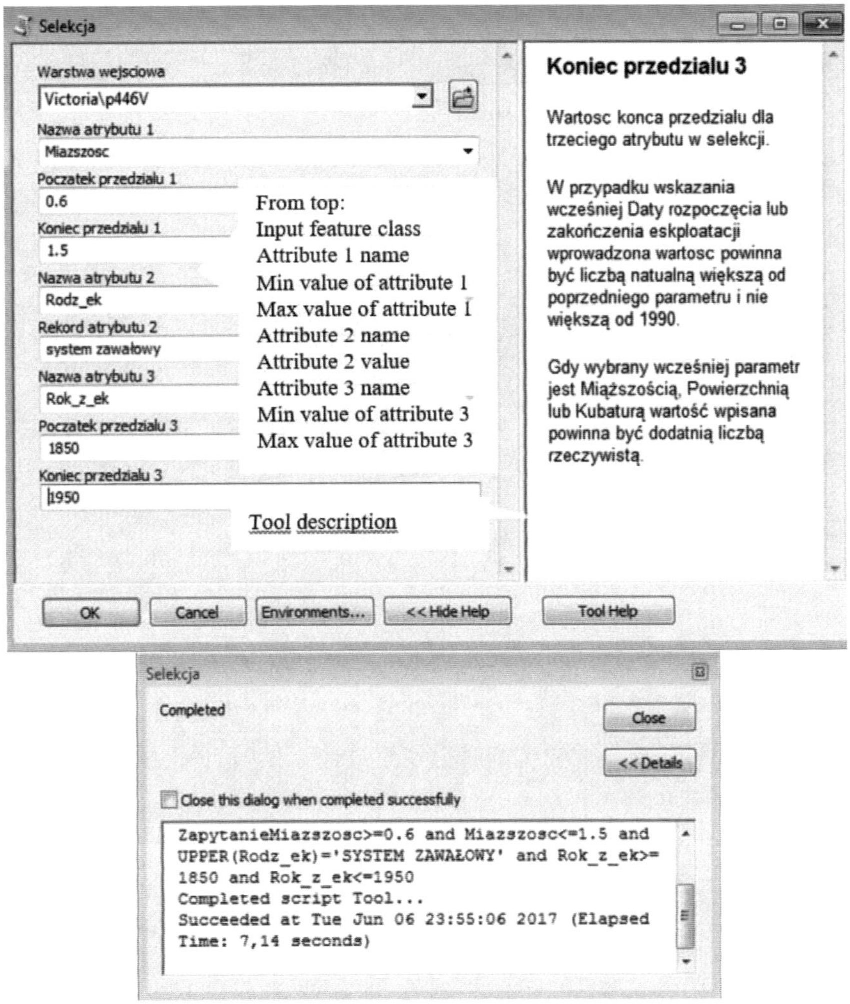

Fig. 3. Graphical user interface of representative tool (top) and result of its application (bottom) [22].

5 Development of web map application

In the development of test web map application for interactive visualisation of spatial and descriptive information on historical mining in Walbrzych open source QGIS (http://www.qgis.org/pl), Geoserver (http://geoserver.org) software distributed under the GNU General Public License have been used. Geospatial data representing particular coal fields and coal parcels in vector format have been stored in PostgreSQL database with PostGIS interface. To browse geospatial data provided by Geoserver the Geoexplorer application has been utilised. Implementation of the database required change of character encoding to UTF-8 from Windows-1250 code and change of attribute column names in data imported to new database for the three mines: *Thorez/Julia*, *Victoria* and *Walbrzych*. Then, the following scheme of naming thematic layers representing coal parcels have been used *Mine_name_Mine_coal_layer_number*. Configuration of the system included creation of workspace with warehouses storing particular thematic layers with assigned workspace, data source, host name and PostgreSQL database name, user name and password.

The last part of web map preparation concerned definition of Styled Layer Descriptor (SLD) file, which is a profile of the Web Map Service (WMS) standard containing user-defined symbolisation allowing for controlling the visual portrayal of geospatial data. In this step unique for each mining system used in Walbrzych mines signature marks have been defined in accordance with mine plans drawing rules. The proposed visualisation of thematic layers representing coal parcels according to mining system used has been presented in Fig. 4.

The GeoExplorer application has been configured with the following main functionalities: zooming in/out, panning, displaying layer properties, changing layer transparency, modifying layer style. In addition, the following tools have been implemented to support basic analyses in web map browser, i.e.: measurement tool, feature information in a pop-up window, attribute queries based on one or more criteria.

These functions have been shown in Fig. 5a and 5b on the example of one coal layer in the Julia/Thorez mine. Fig. 5a shows pop-up window with formatted descriptive information associated with any given coal parcel. Mining system, time of working the given coal parcel (start and end time), thickness, as well as area and cubature are provided for user. In the same figure result of using the measure tool on a given coal parcel has been shown. Whereas, Fig. 5b presents outcome of an attribute query with results shown in the form of database records meeting the given criterion returned in a separate table window. In addition, the web map application allows to view location of coal layers against a given base map as a registered WMS service, for example present day land development shown on orthomosaic.

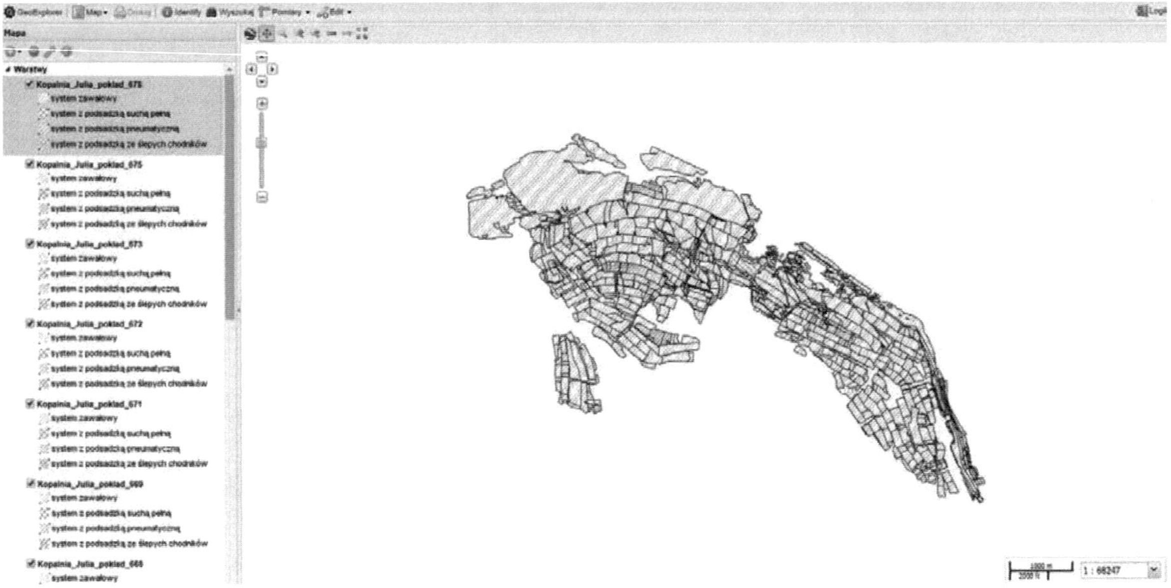

Fig. 4. Location of productive coal layers in the Thorez/Julia mine symbolised according to mining system used presented in open source web map application [22].

https://doi.org/10.1051/e3sconf/20185500002

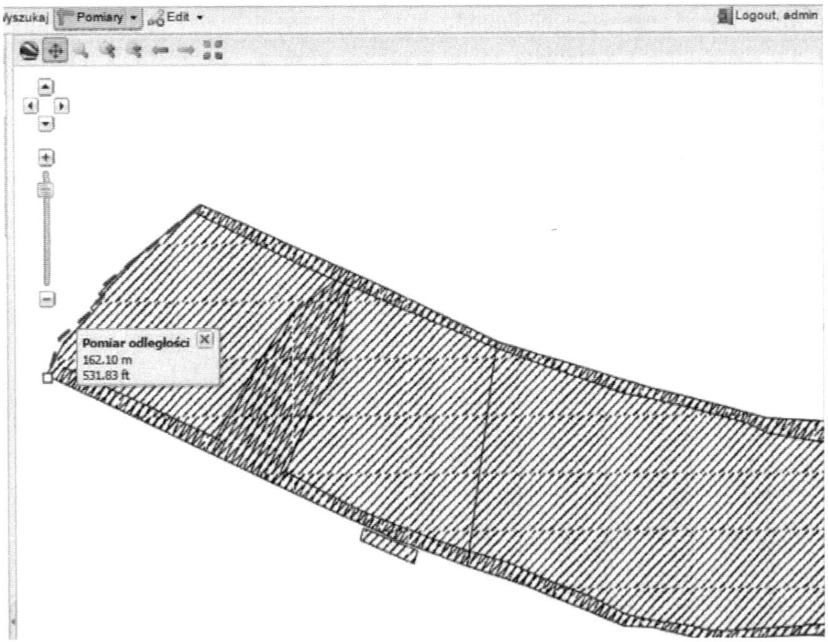

Fig. 5a. Example application of measure and feature information tools.

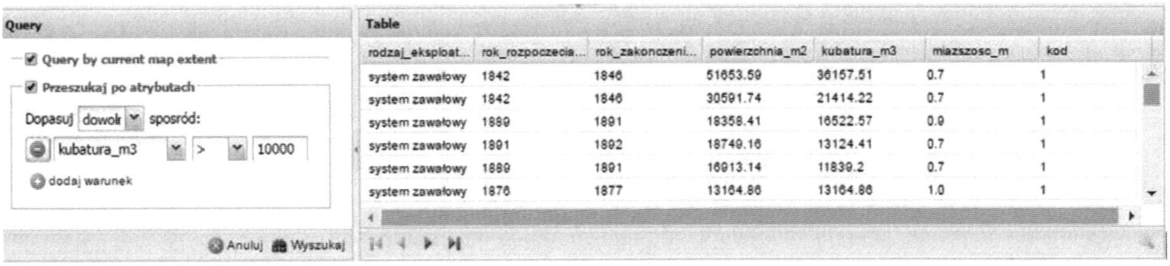

Fig. 5b. Example application of query functionality.

6 Discussion and conclusions

The paper presented universal methodology for development of a GIS and spatial analysis functions assisting studies of underground coal mining on its surroundings. Digitising available data on former mining activity helps to preserve this information for research purposes, e.g. location of former mining sites in the field, as well as create visualisations for dissemination to broader public. An important aspects of such initiatives is the ability to construct complete, in terms of space and time database. This process is sometimes obstructed by incomplete, damaged or even lost data.

Integration of geospatial data representing three adjacent underground hard coal mines in a geodatabase in the presented case, facilitates joint and comprehensive spatial analyses of coal mining and its effects on surface for the entire area of Walbrzych.

With the aim to streamline pre-processing operations necessary for geospatial analyses three query tools have been developed with Python language scripting. These have been used for example in spatial regression analyses connecting deformation factors with the changes observed on the surface (ground movements) [19].

The web map application prepared with open source tools shows how to effectively disseminate geospatial information for a wider audience in an easily accessible way. Web map users can use simple query and spatial analysis tools provided with the application and obtain precise information on location and characteristics of underground mining interesting for local communities such as congruence of coal fields and present day land development.

This work was partly financed from statutory grant no 0401/0123/2017 at the Faculty of Geoengineering, Mining and Geology, Wroclaw University of Science and Technology.

References

1. M. Caro Cuenca, A.J. Hooper, R.F. Hanssen, Journal of Applied Geophysics, **88** (2013)
2. S. Samsonov, N. d'Oreye, B. Smets., International Journal of Applied Earth Observation and Geoinformation, **23** (2013)

3. A. Sowter, L. Bateson, P. Strange, K. Ambrose, M. Fifik Syafiudin, Remote Sensing Letters, **4:10** (2013)
4. L. Bateson, F. Cigna, D. Boon, A. Sowter, International Journal of Applied Earth Observation and Geoinformation, **34** (2015)
5. T. Esaki, I. Djamaluddin, Y. Mitani, Q. J. Eng. Geol. Hydroge., **41** (2008)
6. F. Mancini, F. Stecchi, G. Gabbianelli, Eng. Geol., **109** (2009)
7. K.D. Kim, S. Lee, H.J. Oh, Environ. Geol., **58** (2009)
8. I. Park, J. Lee, L. Saro, Central European Journal of Geosciences, **2** (2014)
9. S. Mei, V. Poncos, C. Froese, EUB/AGS Earth Science Report (2007)
10. M. Mathey, Proceedings of the 13th ISM Congress (2013)
11. J. Fiszer, A. Gogolewska, *Changes of groundwater conditions during the process of Lower Silesian Coal Basin closure* (Wroclaw University of Technology, Institute of Mining Engineering Report No. 35, 2003, in Polish)
12. K. Kominowski, Geology and hydrogeology, [In:] A. Kowalski (Ed.) *Mining and surface protection. Experiences from the Walbrzych mines* (Central Mining Institute, Katowice, 2000, in Polish)
13. S. Kożuchowicz, J. Oprychał, *Geological documentation of coal deposits in the Walbrzych Hard Coal Mine Walbrzych* (Przedsiębiorstwo Geologiczne, Wrocław, 1984, in Polish)
14. J. Blachowski, Mining Science, **X:1** (2008, in Polish)
15. J. Blachowski, P. Stefaniak, Mining Science, **135(42)** (2012, in Polish)
16. J. Blachowski, P. Stefaniak, Mining Science, **20** (2013)
17. J. Blachowski, W. Milczarek, P. Stefaniak, Nat. Hazards Earth Syst. Sci., **14** (2014)
18. J. Blachowski, W. Milczarek, Geological Quarterly, **58:2** (2014)
19. J. Blachowski, Natural Hazards, **84** (2016)
20. W. Milczarek, *Analysis of ground surface changes after mining exploitation in the chosen region of the former Walbrzych basin* (Ph.D. thesis, Wroclaw University of Technology, Poland, 2011, in Polish)
21. J. Blachowski, A. Nowacka, Mining Science, **133(40)** (2011)
22. P. Herkt, *Development of integrated geoinformation system of abandoned underground coal mines in Wałbrzych* (Master thesis, Wroclaw University of Science and Technology, Poland, 2017)

https://doi.org/10.1051/e3sconf/20185500003

A review of methods of cartographic presentation of urban space

Joanna Bac-Bronowicz[1,*], and *Gabriela* Wojciechowska[1]

[1]Wroclaw University of Science and Technology, Faculty of Geoengineering, Mining and Geology, 27 Wyb. Wyspianskiego St., 50-370 Wroclaw, Poland

Abstract. The aim of the article is to present selected solutions applied in urban planning, which, even contemporarily, can be used as models. The paper presents a review of selected concepts of model cities from the antiquity to contemporary times. The article presents methods of creating three-dimensional city models (or their flat presentations in so-called two and a half dimensions), including the historical conceptualization, portraying both the practical and aesthetical features of former map projections. Referring to historical examples, the article also includes historical and contemporary urban 3D models corresponding to analyzed conceptions.

1 Introduction

Portraying of the surrounding reality has been an immanent element of our civilization since its beginnings. Since the prehistory, depending on the predominant aesthetic movement, the imagery of surrounding environment was presented through the means of cave paintings, rock reliefs (Fig. 1), as well as paintings in interiors of temples and other structures.

Fig. 1. Forma Urbis Romae, approx. 200 AD [1].

Maps serve numerous important functions, such as: historical, social, cultural, educational, directional, cognitive, operational and guiding. The guiding function is also an important property of town maps and tourist maps [2]. Graphic notation has been serving people since the beginnings in order to create holistic spatial imageries which enable them not only the identification of landmarks and their spatial layouts, but also acquiring the knowledge concerning the unknown places, both existing and artificial, through the means of analogy [2].

Cartographic methods have developed significantly over the recent years. Thanks to the use of computer technology, it seems that the possibilities of precise cartographic representation are endless. Changes in remote sensing made it possible to collect increasingly accurate 3D data concerning the environment, as well as information about new products and services in 4D. It is believed that a number of changes in cartography during the next five to ten years will be caused by 4D cartographic products [3].

It is believed that the increase of urbanization will present a global challenge for cartography. The analysis of data concerning the course of global urbanization reveals that the population in cities increases threefold quicker than the average global population. The United Nations estimate that, by the year 2050, the global population will amount to over 9 billion, of which 80% is estimated to live in urban settlements. As a result, the notion of Smart City was coined. In general, it may be assumed that the above term refers to a certain idea of style and functioning of a city, or its management described in the language of modern technologies. Problematic aspects of smart cities are both multi-faceted and multi-dimensional, as they incorporate humanistic, social, environmental and technological problems [4]. Intelligent systems will be used in order to develop new ways of presenting the complex spatial relations to their observers. The development of augmented and virtual reality will enable people to work with data in a new way. The integration of smart technologies and efficient city management models for sure will require quick, automated digital cartographic models with possibly highest data quality.

Representations of cities will cease to include merely presentations of external development. Current techniques used for analysis, modelling and presentation of geoinformation become increasingly improved and enable creation of virtual city models aiming at management of urban infrastructure. Apart from praising

* Corresponding author: joanna.bac-bronowicz@pwr.edu.pl

the infinite number of current possibilities, it is also important to take notice of changes in cartographic presentations of cities which occurred during the history.

Among numerous cartographic presentations those which present projections crucial for the development of the history of cartography have been selected, reflecting the character of a given period. When analyzing historical reports, it is worth noticing that, despite the location and time period, city plans have always had an important cultural aspect. By presenting the available knowledge concerning the surroundings, maps raised and increased awareness of many societies. Moreover, maps are a necessary element in gaining knowledge concerning the surroundings and presenting projects related to spatial development.

The aim of the article is to present selected solutions applied in creation of city maps, which might serve as models even in the contemporary day and age.

2 Urban cartographic representations – from antiquity to contemporary times

The following part of the paper presents selected examples of cartographic representations of cities, which illustrate the most prominent art movements which influenced the development of this field of knowledge, as well as those which are considered as groundbreaking in the course of history of cartography. It commonly occurs that the same discoveries are made several times during the course of history. Some inventions become forgotten due to the collapse of civilizations, just to be rediscovered in the following ages. We are aware that metric plans were known to certain ancient civilizations, it is also commonly known that during the Medieval Period they were forgotten only to be rediscovered during the times of the Renaissance and, until today, considered as basic cartographic representations of cities.

The development of cartographic maps and methods of representations of cities were connected with the perception of reality and artistic practices characteristic for a given epoch. Aspective art, which was also known in Bronze Age, was often used by, among others, ancient Egyptians, who depicted three-dimensional objects by combining their views from above and from the side. Such representations were commonly used in Medieval maps, e.g. the map of Jerusalem from 1120 from the chronicle by Robert of Saint-Remy [5] and Mappa mundi from Ebstorf (Fig. 2) – where one plane combines a group of communications routes and depictions of building facades. It is widely believed that not many city maps were created in Medieval Europe. Regular use of maps and urban plans was limited to very few specific areas. The idea of making a draft of a map in order to showcase topological and topographical relations (roads connecting particular locations, layout of fields, order of houses along a street) was not popular in the Middle Ages. Medieval maps of small areas and regions were in its time quite unusual realizations, oftentimes presenting substantial originality and ingenuity of the author who might have never seen a map drawn up by someone else. In the Middle Ages it was common to present topographic relations in writing, therefore, descriptions were preferred over maps [6].

Fig. 2. Part of the map from Ebstorf, approx. 1300 [7].

The second half of 13th century is the time of considerable advancements in representation of space and search of a logical conception of object depiction. Former perspective primitiveness is gradually replaced by a type of parallel perspective, often unhampered by the rules of geometric representations, in order to evolutionally transform into linear perspective (the elementary principles of its creation were determined already in 15th century) [8].

Parallel perspective (used in the art of Far East and ancient Greece) uses diagonal presentation in order to depict structures spatially; both the façade and depth are visible. Buildings which are further away are placed higher than closer ones, however their size is not changed. According to W. Strzemiński, visual awareness is conditioned by the process of historic transformations. Linear perspective, which is currently considered as traditional perspective and which corresponds to natural human vision, was oftentimes rejected in ancient Greece. Plato considered painters who created using the said technique as creators of illusions and liars [9]. It was a new perspective which surprised his contemporaries as it did not comply to their visual awareness.

The Paris map (Fig. 3) published by Olivier Truschet and Germain Hoyau in mid-16th century [12], is an interesting instance of urban cartographic representation created with the use of parallel perspective. The map contains schematically presented buildings together with ensuring the visibility of streets and their names, as well as information concerning places of importance for the inhabitants, i.e. the location of the place of ball games or trail of upstream barge haul. Former depictions of cities are invaluable sources of information concerning their development. They may serve as source material for the historical reconstructions of urban planning schemes, i.e. Fig. 4., which presents the reconstruction of 14th century Paris with the view of la Sainte Chapelle, made by Dassault Systèmes. Admittedly it is a reconstruction of a period earlier then the aforementioned map, nevertheless, it presents the current possibilities of virtual reconstruction. Reproduction of cartographic legacy in a digital form grants a new possibility of maintaining and examining historical cartography. Digital techniques currently applied in cartography enable a georeferential analysis of historical maps.

Combining of the knowledge of cartographers, architects, historians and urban planners makes it possible to reconstruct the historical layout of an urban settlement together with attributes of reconstructed facilities. An example of such works is the first urban map from the 19th century Gregorian cadastre in Bolonia [13]. In the above study the researches used GIS tools as well as 3D modelling and historic sources in order to analyze the image of the city from the perspective of proprietorship of its inhabitants. The analysis of 19th-century layout of the city was conducted after the conversion of information into a georeferential 3D model enabling synthetic analysis of ownership aspects, which, prior to the analysis, were not noticeable during the direct inspection of documents. An additional advantage of the model is digitalization of original documents, which were desired to be accessed by researchers.

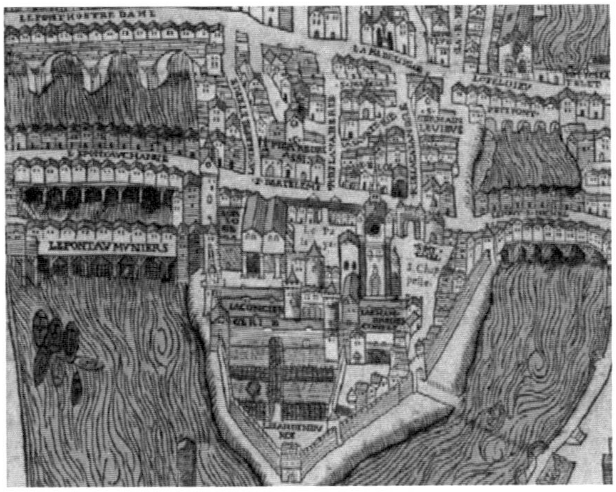

Fig. 3. Part of the Paris map from approx. 1550 [10].

Fig. 4. Reconstruction of 14th century Paris by Dassault Systèmes (the view of la Sainte Chapelle) [11].

Together with the development of the theory of perspective, the representations of cities became increasingly common. One of the milestones in 3D cartography is the map of Venice (Fig. 5) from the year 1500 created by Jacop de Barbari, the apprentice of Albrecht Dürer [14], which is one of the first bird's eye view representations of the entire city in history. The map was depicted on six woodcuts; its precision, together with the fact that many of the buildings from that period still remain intact, enable touring the city with the use of the said map even today. Fig. 6. presents the contemporary outlook of Venice obtained with the use of satellites, published by Google Earth portal. The comparison of both depictions shows the mastery of creators of the Renaissance map, who were able to capture the reality using only basic tools and their imagination. The accuracy of this historic representation (Fig. 7) is so high that it is easily possible to find characteristic facilities and examine the changes which took place in the historic development over last 500 years.

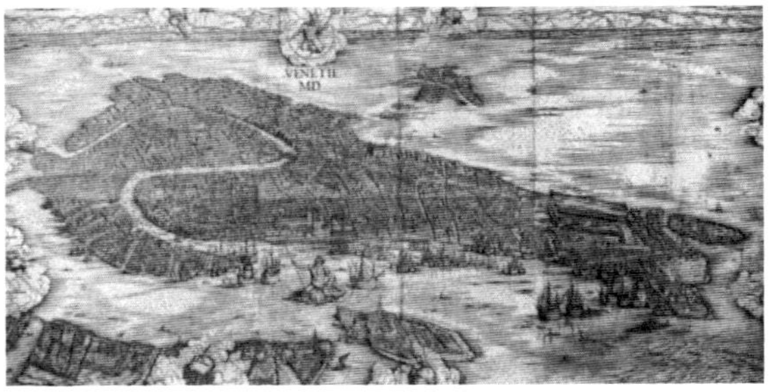

Fig. 5. Map of Venice by Jacopo de Barbari (approx. 1500 r.) [15].

Fig. 6. Venice in 3D, Google Earth [16].

City maps enabled not only to localize places in the surrounding space and manage the area, but they also were symbols of power and dominance of a particular urban settlement.

Together with the development of illustration techniques and measurement methods the maps became increasingly perfected. There is a strong relation between art and cartography, as, especially during the period of the Renaissance, painters were often authors and coauthors of maps (A. Dürer, L. da Vinci, J. van Eyck).

Fig. 7. Basillica Santa Maria Gloriosa dei Frari represented on the map of Venice by Barbari and a 3D view from Google Earth [15, 16].

The map of Milan by Leonardo da Vinci is considered to be a breakthrough in the history of cartography (Fig. 8), as it is believed to be the first cartographic attempt to transfer the observations from the ground level to the bird's eye view with sheer use of artist's imagination, which was in opposition to the contemporary representations of cities painted from a specific point as perspective views. Da Vinci drew the three-dimensional view of Milan using the principles of mathematical perspective basing on the obtained projection [17]. In his latter cartographic works, da Vinci continued creating two-dimensional map representations, such as those known in modern cartography, rejecting the bird's eye view, i.a. Imola town map. Two-dimensional cartographic representations of cities were also created by Alberti, who was the author of the map of Rome, which unfortunately was not preserved and is known only from its descriptions [17].

Former maps often contained representations of urban panoramic views. By the end of 15th century flamboyant maps and urban outlooks became popularized in royal, papal and aristocratic palaces, where they served both informative and decorative function. Urban views were the most popular cartographic images in the Renaissance and their popularity was so high that they were often encountered in such publications as Braun and Hogenberg's Civitates orbis terrarum and used for interior design. They were deliberately placed in the suites intended for prominent guests in order to impress them with representations of surrounding areas. Since the second half of 15th century urban views appeared also as painted cycles, which gained even more popularity. The meaning and functions of painted cycles with the city view did not differ significantly from the cycles of maps, although they constitute a separate group [19].

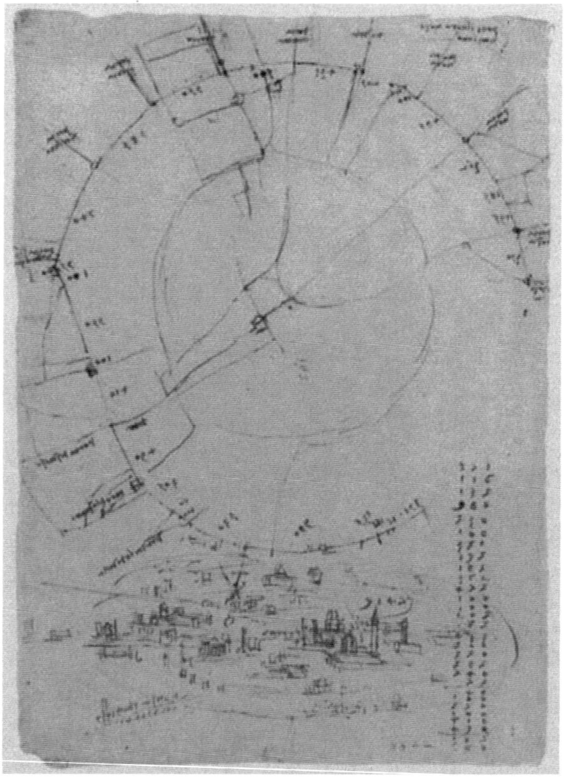

Fig. 8. Plan and view of Milan, Codex Atlanticus (1478-1519) [18].

Renaissance cartography features maps of so-called ideal cities – which served as examples of the search for perfect dwelling space. The design assumptions applied in those works emphasized the aesthetics and symmetry, which is clearly visible in cartographic representations of those layouts. The Renaissance brought about change in the perception of a city. Since 15th century, city has become an observed and analyzed phenomenon. Together with its studies new theories concerning the ideal city form were developed. It was also the time of perfection of presentation – the Renaissance marks the beginning of the use of perspective. The Renaissance space was highly ordered and unified, shaped according to rigorously applied proportions [20]. In 16th century numerous model cities were planned, for instance, functioning until this day, model city, star fort – Palmanova, Italy (1593) by Vincenzo Scamozzi and Zamość, nicknamed as the Pearl of the Renaissance and "Padua of the North". Zamość, located in the eastern part of Poland, is a realization of a Renaissance model city designed by an Italian architect, Bernardo Morando. Its construction was commenced in 1579 and it was commissioned by Jan Zamoyski, who wanted to build the headquarters and an impregnable fortress for his family, which would stun with its architecture. The town

https://doi.org/10.1051/e3sconf/20185500003

was included in the UNESCO World Heritage List in 1992 (Figs. 9, 10).

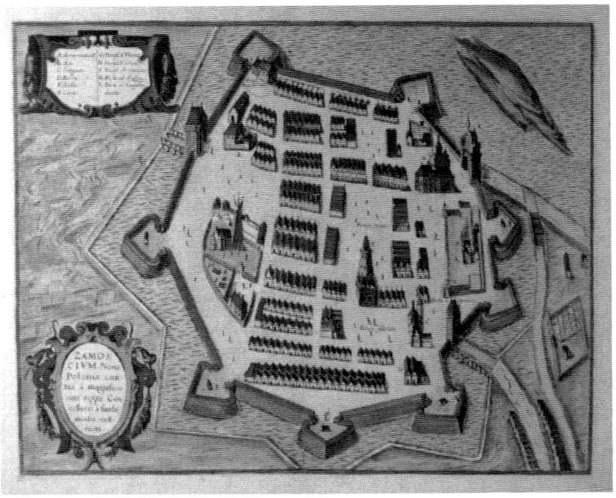

Fig. 9. Plan of Zamość by Georg Braun and Franz Hogenberg, 1617 [21].

Fig. 10. Contemporary view of Zamość, fot. M. Poznański [22].

It is also worth noticing that old maps were created in various purposes – in some cases the way of graphic presentation was strongly dependent on the purpose and function of the map. It is also crucial to mention the use of intentional perspective, which considered the hierarchy of depicted structures in the pictorial representations – more important objects were painted bigger. An example of a map with a purposeful distortion of the imagery may be the map of Moscow from 1610, which currently would be qualified as an anamorphic map. Throughout centuries, the anamorphic conceptualizations were a natural way of notation and understandable form of transfer of information, which is evidenced by: drawings of trade routes, Roman operational military maps, pilgrim maps, the above-mentioned Polish "spy" map of Moscow (Fig. 11), or an 18th-century draft of contemporary dioceses of north Poland. Violation of the Euclid rules in case of the aforementioned models resulted in the increase of their functional qualities e.g. for the troops besieging Moscow it was the courses of trackways from the gates of the city towards the Kremlin, which were accentuated due to purposeful extending [2]. The actual course of trackways is presented on the Moscow map from 1597 (Fig. 12).

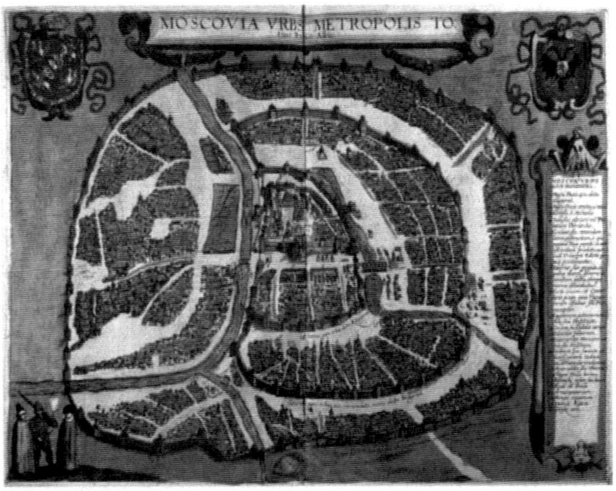

Fig. 11. Map of Moscow, 1610 [23].

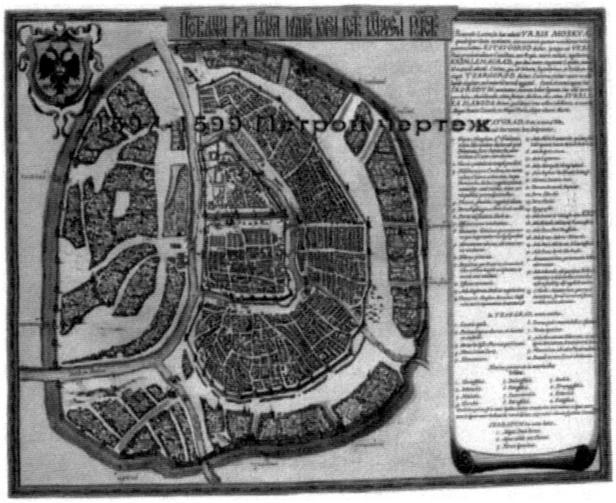

Fig. 12. Map of Moscow, 1597 r. [24].

Together with the development of cartography, conventional markings replaced pictorial images. Gradual division between cartography and art commenced with the introduction of new tools and techniques of measurement, which began to replace works of artists as early as in 15th century. What is more, even after the establishment of new scientific measurements in 17th century, the techniques of landscape sketching still maintained a direct relation between cartography an art and in doing so remained useful for natural scientists, engineers and soldiers [17].

Popular projections on 17th-century maps drawn up basing on direct measurements were urban dioramas, sometimes even urban plans, as well as chorographic presentations in the form of imponderables, which oftentimes artfully combined science and art. Such way of presentation of urban landscape may be observed, for instance, on a map of the Mediterranean Sea created by Pierre Mortier in Amsterdam in 1694.

18th century is the period of Enlightenment, which might be called as the era of maps. The concept of a geographic map was fascinating for the contemporary, as it embodied the complex discipline of "mathematical cosmography". Governmental and administrational institutions became to rely more strongly on maps in order to regulate and control their territories. Developing printing and visual culture led to creation of maps both in the form of manuscripts and prints, such representations were created in accordance with common aesthetics of layout and design [25]. Graphic representations oftentimes were lacking in "unnecessary decorative elements". Conventional design and presentation standards influenced also topographic and other big-scale maps. That period is also the time of creation of city maps (2D) with high accuracy and detailing of presentation together with high quality of graphics. Fig. 13 presents original cartographic presentation from 1705 depicting generalized high-density city blocks in Paris from the period of 1422-1589.

Fig. 13. Fragment of plan of Paris in 1589 (from 1705), the sixth of eight chronological maps of Paris from Traité de la police by N. de La Mare [26].

19th century is the age of cartography and popularization of thematic maps. Creation of maps became institutionalized, specialized and professionalized. The need for maps was constantly increasing – as a result of the development of tourism, mass education and introduction of cheaper printing techniques, as well as creation of new urban infrastructures, which led to spreading of cartographical knowledge and development of corporate producers of maps. Industrialization influenced the aesthetics of map design, partially determined by new printing technologies and introduction of color printing. One of the most well-known urban maps from that period is *Descriptive Map of London Poverty* from 1899 by Ch. Booth depicting the complexity of social geography of the city [Fig. 14]. The degree of wealth of inhabitants was marked using a colored scale.

20th century marked a breakthrough in the history of maps. The transition from the paper to digital format resulted in incredible changes as it commenced the era of dynamic and interactive maps [28]. Systems of geographic information radically changed cartographic institutions, while satellite positioning and mobile communication revolutionized the way of searching for information.

Fig. 14. Fragment of Ch. Booth poverty map of London [27].

Currently, it is very difficult to imagine spatial planning, crisis management or investment design without the use of spatial 3D data. In order to create such spatial models, various measurement techniques are applied, such as: data from terrestrial and aerial laser scanning, digital terrain model (DTM), digital land cover model (DLM), aerial photographs and images taken by unmanned aircraft, satellites and radars. In many countries 3D models of entire cities are created basing on such spatial data, an example of which may be a project by Polish Head Office of Geodesy and Cartography, which will include 3D models of buildings located throughout Poland, as well as devising tools which enable access to the information using an Internet browser. International standard CityGML 2.0, issued by Open Geospatial Consortium, was selected as the format to record 3D building models. All models will be designed with the LoD 2 [29]. Basing on the prior experience of numerous countries, it can be unequivocally stated that 3D e-services are and will continue to be widely applied in various types of analyses, such as, e.g.: visualizations applied in navigation, urban planning, visibility and shading analyses, measuring of terrain height and its solar potential, facility management, identification of aircraft obstacles, creation of thematic maps, as well as simulations and modelling of various phenomena, such as, rescuing missions or noise emission in urban settlements [4, 29-31]. What is also worth mentioning is the possibility of using 3D data in the inventory processes. Three-dimensional building models are used in spatial planning as well as architecture and urban planning. It is worth noticing that 3D data an e-services are also sources used by experts in promotion and marketing of towns and regions. The abovementioned examples of application of such data serve as proof of its immense potential. Depending on expected results, 3D models might be texture mapped using such techniques as photogrammetry and laser scanning [32].

There has been intensive ongoing work on technologies enabling creation of fully virtual environments, which would integrate the elements of physical with virtual activities (augmented reality). Soon, the screen used as the means of sharing information might be replaced by three-dimensional space. There are numerous tools, as well as open-source software, which might be used to display various cartographic representations, e.g. Google, as well as software tools and technologies enabling presentation of statistical data on digital maps in a browser, e.g. solutions provided by Google [33]. The applications are interactive and might be quickly updated, moreover, their content might be easily changed using available libraries and own models created with the use of such tools as SketchUp software, which later may be shared in Google Earth, a program which contains numerous detailed representations of popular structures and sights from all over the World.

3 Summary

When analyzing city maps the way of arranging facilities on them may be simplified with the use of the following taxonomy: symbolism in antiquity and Middle Ages, transition towards realism in the Renaissance and later gradual distinction between cartographic representations and painting imagery. At that time two-dimensional representations were dominant in cartography and the standardization of symbols for structures viewed from above was commenced. The end of 18th century is the time when symbols were commonly used instead of images, which led to losses in decorativeness together with the increase in transparency and precision [17]. From the beginning of 17th century, city maps and urban panoramas were often presented in the form of imponderables, which harmoniously combined art and science. During the period of Enlightenment maps began to have more scientific character and their authors resigned from "unnecessary decorative elements". 19th century marks the era of cartography, popularization of thematic maps and a pivot towards geoinformation – precise location of structures. The representations were mainly two-dimensional, with satisfying accuracy and in accordance to the requirements of the representation scale.

20th century is a breakthrough in the history of maps. Transition from paper to the digital format, as well as systems of geographic information led to immense changes in cartography.

First interactive maps were not always created in accordance with cartographic good practices. The same issue occurs with automatically created 3D city models. Their representations are accurate in terms of geometry, nevertheless, they do not always meet the requirements of correct perception, due to a big number of details, selection of colors, textures, descriptions and map symbolization. 3D city models may have a wide application, that is why, in order to present them effectively, it is also necessary to bare in mind the quality of their visual representation. The authors suggest that one of the solutions which would improve the quality of visual presentation would be the inclusion of graphic designers in the designing process.

It is also necessary to bare in mind the issue of understanding and experiencing space through the use of new technologies. Virtual reality is not a copy of real urban area due to perception differences. In analyses of spatial models which map real structures other cognitive relations occur, which change our visual perception of, e.g. the body of the structure. As an example of such phenomenon might serve the study conducted by researchers from Bío-Bío University in Concepción, Chile, whose results led to the conclusion that in virtual space notions such as neighborhood and distance lose their importance and that intangibility of virtual space significantly weakens the perception of the organization of form [34-35]. In the view of the above, studies which strongly rely on such features as: visibility, assessment of localization and aesthetics, for instance urban analyses, when conducted basing on digital reality, may lead to faulty conclusions if the differences in perception are not acknowledged.

In classically managed cities two causes of insufficient use of geoinformation were noticeable: lack of appropriate resources of spatial data and, so called, silo (vertical, trade) urban resources management, including spatial data [4]. As a result of the development of new technologies, the technical ways of 3D and 4D presentation will have to be developed simultaneously with the advances in data processing and the increased accessibility of spatial data of various quality will be challenging for its users. A novelty in the attitude towards previously used representations is the necessity of evaluation of the quality of data and taking responsibility for newly created city models, which might be the basis for multiple and various decisions stemming from the Smart City philosophy. It does not mean that we have to abandon our imagination, moreover, it does not seem like this would happen anyway. Artists create their own visions, an example of which might be a popular collection of graphics entitled "GRAPHIC EUROPE an Alternative Guide to 31 European Cities" [36], which presents a return of emphasizing places important to the author with the means of drawing. There is also a plethora of visions presenting city development, such as this of Barcelona [37], or the design by Kevin Schopfer, who seems to have thought of everything in his NOAH (New Orleans Arcology Habitat) project, where he designed a railway and elevator network destined for the communication. Apart from standard functions, in the city there were also: several hotels, a casino and various other facilities [38]. Because, as Arthur C. Clarke said, "Any sufficiently advanced technology is indistinguishable from magic".

References

1. Realm of History, https://www.realmofhistory.com, access 12.06.2016
2. E. Krzywicka-Blum, *Map Functions* (Springer International Publishing, Switzerland, 2017)
3. UN-GGIM, United Nations Initiative on Global Geospatial Information Management. Future trends in geospatial information management: the five to ten year vision, http://ggim.un.org, access 10.06.2017
4. D. Gotlib, R. Olszewski, *Smart City. Informacja przestrzenna w zarządzaniu inteligentnym miastem* (Państwowe Wydawnictwo Naukowe, Warszawa, 2016)
5. J.A. Wendt, *Skarby kartografii* (Wydawnictwo Arkady, Warszawa, 2013)
6. P.D.A. Harvey, [In:] D. Woodward, J.B. Harley (Eds.) *The History of Cartography vol. 1* (University of Chicago Press, Chicago and London, 1987)
7. Wikimedia Commons, https://upload.wikimedia.org, access 02.06.2017
8. K. Bartel, *Perspektywa malarska, tom II* (Państwowe Wydawnictwo Naukowe, Warszawa, 1958)
9. W. Strzemiński, *Teoria Widzenia* (Wydawnictwo Literackie, Kraków, 1958)
10. Wikimedia Commons, https://upload.wikimedia.org, access 15.06.2017
11. Dassault Systèmes, https://www.3ds.com, access 15.06.2016
12. M. Pelletier, [In:] D. Woodward (Ed.) *The History of Cartography vol. 3* (University of Chicago Press, Chicago and London, 2007)
13. G. Gatta, E. Arioti, G. Bitelli, Journal of Cultural Heritage **23**, 68-76 (2017)
14. J. Brotton, *Słynne mapy. Arcydzieła światowej kartografii i ich tajemnice* (Wydawnictwo Arkady, Warszawa, 2016)
15. Wikimedia Commons, https://upload.wikimedia.org, access 02.06.2017
16. Google Earth, https://www.google.com, access 02.06.2017
17. R. Rees, Geographical Review **70**, 1, 60-78 (1980)
18. C. Moffatt, Leonardo da Vinci and mapmaking, http://faculty.virginia.edu, access 20.06.2017
19. F. Fiorani, [In:] D. Woodward (Ed.) *The History of Cartography vol. 3* (University of Chicago Press, Chicago and London, 2007)
20. E. Szpakowska, Przestrzeń i Forma **16**, 121-154 (2011)
21. Wikimedia Commons, https://upload.wikimedia.org, access 27.07.2017
22. M. Poznański, Stare Miasto - Zamość, http://lublin.wyborcza.pl, access 10.10.2017
23. Wikimedia Commons, https://upload.wikimedia.org, access 17.06.2017
24. Moscow Plans, http://abodye.ru, access 01.07.2017
25. M.H. Edney, M. Pedley (Eds.) *Cartography in the European Enlightenment Volume 4* (The University of Chicago Press, 2018)
26. Wikimedia Commons, https://upload.wikimedia.org, access 16.03.2017
27. Wikimedia Commons, https://upload.wikimedia.org, access 15.03.2017
28. M. Monmonier, *Cartography in the Twentieth Century* (University of Chicago Press, 2015)
29. Zespół ds. 3D w projekcie CAPAP w GUGiK, Geodeta **9**, 22-25 (2017)
30. M. Konecny, S. Zlatanova, T.L. Bandrova, *Geographic Information and Cartography for Risk and Crisis Management* (Springer-Verlag Berlin Heidelberg, 2010)
31. F. Biljecki, J. Stoter, H. Ledoux, S. Zlatanova, A. Çöltekin, ISPRS Int. J. Geo-Inf. **4**, 2842-2889 (2015)
32. Ch. Frueh, R. Sammon, A. Zakhor, *Proceedings of the 2nd International Symposium on 3D Data Processing, Visualization, and Transmission* (2004)
33. K. Król, GLL **2**, 91-106 (2016)
34. R. Zapłata, Quart **1-2**, 124-137 (2017)
35. A. Asanowicz, Architecturae et Atribus **4**, 5-12 (2012)
36. Z. Hanaor, *GRAPHIC EUROPE an Alternative Guide to 31 European Cities* (Cicada Books Limited, London, 2009)
37. Kaspersky, Earth 2050, http://adsoftheworld.com, access 27.07.2017
38. E.K. Schopfer, NOAH (New Orleans Arcology Habitat), http://homedesignmini.blogspot.com, access 27.07.2017

Detection of surface subsidence using SAR SENTINEL 1A imagery and the DInSAR method – a case study of the Belchatow open pit mine, Central Poland

Jaroslaw Wajs[1,*], and *Wojciech J.* Milczarek[1]

[1]Wrocław University of Science and Technology, Faculty of Geoengineering, Mining and Geology, 27 Wyb. Wyspiańskiego St., 50-370 Wrocław, Poland

Abstract. Open pit mining hazards monitoring is a crucial task. The paper focuses on the first analysis of surface subsidence in the open pit mining area obtained by Synthetic Aperture Radar active remote sensing technique. Presented methodology of Differential Interferometric SAR provides an excellent data source for mining vertical displacements monitoring. The study was performed in the Belchatow open pit lignite mine in Poland, Central Europe. The SAR SLC imageries acquired by the SENTINEL 1A satellite for the 124-descending track in two periods, between 03.10.2016–15.10.2016 and 15.10.2016–27.10.2016 have been used in the analysis. The post-proceed satellite line of sight (LOS) displacement indicates vertical changes of the surface within the dumping and excavation area. The analyzed region of interests (ROI) shows total subsidence of ca. -67mm, whereas the excavation areas show a trend of terrain uplift during the analyzed periods.

1 Introduction

Monitoring of mining areas with application of land survey methods is the fundamental task of mine surveyors. When performed with the use of classical survey methods, these tasks result in point sampling of the reality by geodetic observation lines and surface sampling using the raster matrix in the analyzed mining area. Three-dimensional modelling of open cast mines determines the spatial coordinates of characteristic terrain objects. Classical measurement methods in mining surveys require time-consuming fieldwork and are loaded with errors of the surveyor related e.g. with ground sampling. The result of the performed investigations is XY coordinates and their elevation values in an accepted reference system [1]. Fig. 1 presents the comparative analysis of selected active remote sensing sensors as sources of spatial data. This analysis was prepared based on compilation [2]. Remote sensing may be an alternative data source for the spatial database of the open pit lignite mine. Since analyzed object is an open pit mine, the obtained map of terrain subsidence has to base on high temporal, spatial data. These features are characteristic for the sensor Sentinel 1A which is present-day operational and open access.

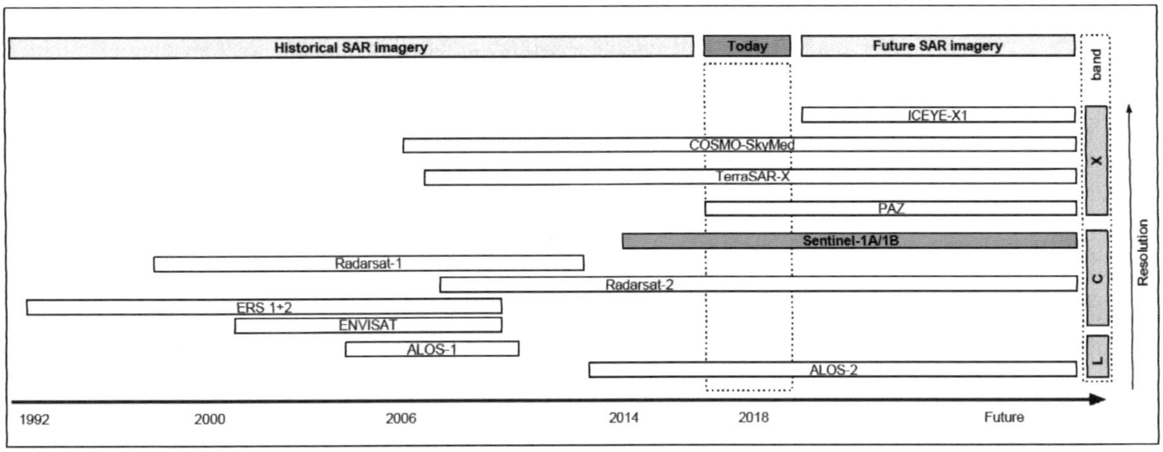

Fig. 1. Historical and present-day active Remote Sensing SAR sensors.

* Corresponding author: jaroslaw.wajs@pwr.edu.pl

2 Motivation

Open pit lignite mine is an object whose topography changes over time. One of the most important information about this anthropogenic object are vertical displacements. Due to the fact, that the remote sensing allows to acquire high spatial and temporal data, the frequent information about object of interests can be provided. With the development of remote sensing missions of the European Space Agency (ESA), remote sensing data have become a perfect source of information on the state of objects in macro scale. In November 2012 Poland became an ESA member and at the same time a beneficiary of the GMES (Global Monitoring for Environment and Security) programme. The continuation of the GMES is the Copernicus programme, in the frame of which SENTINEL satellites were introduced. The constellation of Sentinel 1A/1B satellites is a source of terrain imageries with high temporal and geometrical resolution accomplished due to specially applied precise orbits, whose precision was determined in 3D at the level of 5 cm in [3]. The instrument installed on the Sentinel 1 satellite is a Synthetic Aperture Radar operating in band C with 5.65 cm wavelength. These components allowed for application of SAR data in the monitoring of the study object. The active SAR instrument allows to deliver the detailed and accurate information about ground subsidence. Over the past two decades, InSAR techniques have evolved from the use of differential interferograms (DInSAR) in the late 1990s [4, 5] to Persistent Scatterers Interferometry (PSI) in the 21st century [6, 7]. Monitoring of post mining settings was presented by [8] and [9]. There are many methods of SAR data processing, which, along with the development of high resolution SAR satellites such as TerraSAR-X observing in band X, allow for mine monitoring with high spatial resolution and precision. At first, reports on SAR data processing covered urban agglomerations [10] and underground mines, where stable PSI diffusers are relatively easy to acquire [11, 12]. PSI in fact overcomes limitations related to the atmospheric contributions affecting the standard DInSAR technique and allows millimeter precision to be reached. Monitoring of open pit mines with DInSAR techniques was presented already by [13, 14]. An open pit mine is a dynamic object that changes its geometry with time. Currently, advanced methods of SAR time series high spatial resolution and frequent temporary acquisition give the opportunity to detect changes from space also in open pit mines [15, 16].

3 Study area

The analyses were performed in the PGE KWB Belchatow open pit lignite mine located in central Poland, Europe. The KWB Belchatow object comprising two excavation areas named as Belchatow and Szczercow which are presented in the portion of the Sentinel2A false color imagery acquired on 25.05.2016 (Fig. 2). The analyzed region of interests (ROI) overlapped the mining area and the remote sensing data capacity was reduced to this region. The outline of the open pit and the internal dumping site is marked with a yellow rectangle and marked as A. Areas representing the second excavation Szczercow is depicted by B. The ROI including the Szczercow external dumping site and marked as C was analyzed for possible pseudo-vertical displacements of the terrain. Since analyzed object is an open cast lignite mine, the obtained terrain subsidence is an essential information of mining area.

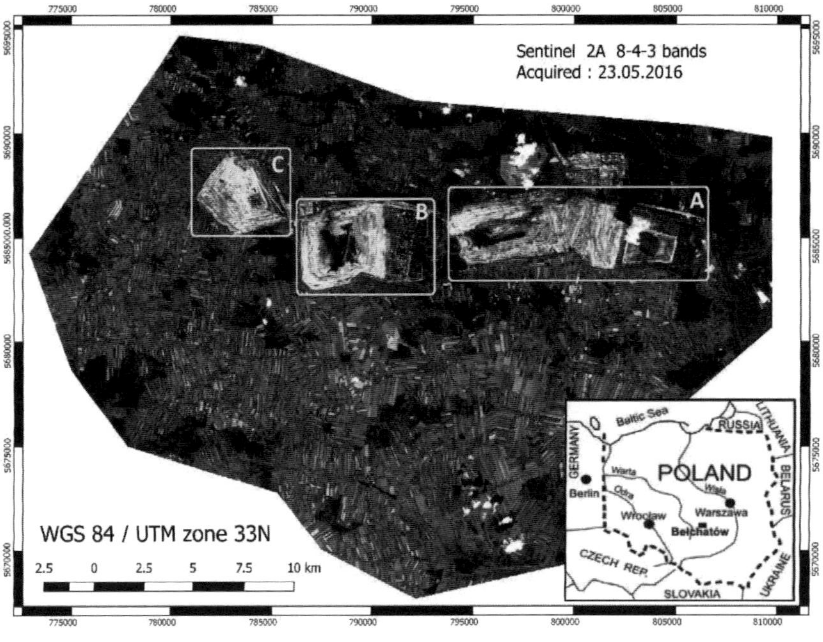

Fig. 2. Portion of Sentinel imagery of the Belchatow open pit lignite mine. The combination is in a false colour 8-4-3 band acquired on 25.05.2016.

Table 1. The visualization of Sentinel 1A 124-orbit SLC IW1 product is presented at Fig. 3.

Sentinel 1A Scene	Date of acquisition [date]	Orbits	Incident Angle [DEG]	Baseline [m]	Temporary resolution [days]
SLC IW1	03.10.2016	124 descending	30.86-36.59	43	12
SLC IW1	**15.10.2016**	**124 descending**	30.86-36.59		
SLC IW1	27.10.2016	124 descending	30.86-36.59	9	12

4 Methodology

4.1 Sentinel 1A data preprocessing

Data subject to analysis were collected from ESA on the SentinelHUB platform (URL: https://scihub.copernicus.eu/dhus). The processed imagery was obtained by Sentinel 1A on 03.10.2016, 15.10.2016 and 27.10.2016 in a Interferometric Wide Swatch, Single Look Complex standard with VV polarization. Additionally, POD Precise Orbit Ephemerides (URL: https://qc.sentinel1.eo.esa.int) were also collected for these imageries. Details are presented in Table 1.

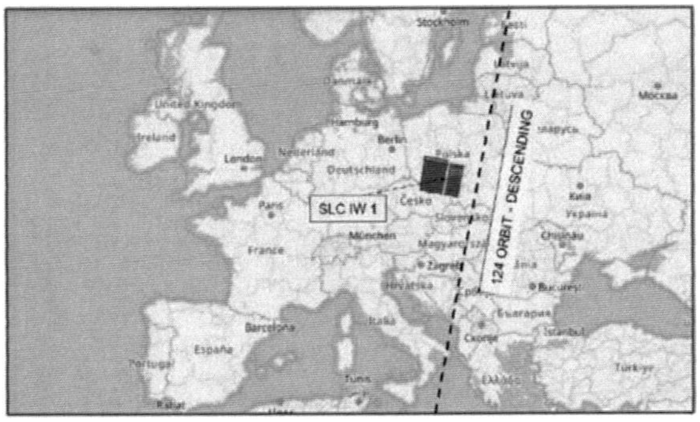

Fig. 3. Visualization of Sentinel 1A 124-orbit SLC IW1 product.

4.2 DInSAR method description

Data from the Synthetic Aperture Radar (SAR) of the SENTINEL 1A mission are characterized by high spatial (5x15 m) and temporal resolution (12 days). Additionally, precise orbits allow the satellite position to be stated precisely and the spatial database B presented in Fig. 4a to be determined. At maintaining optimal construction requirements, a 3D reconstruction of the terrain surface is possible, for example in the SRTM mission. The methodology of SAR data processing presented herein is based on Differential SAR Interferometry (DInSAR) (Fig. 4b). It processes a minimum of two SAR imageries collected in time t1 and t2, respectively, and including the Digital Elevation Model component [18]. The main requirement for accurate processing in DInSAR methodology is to have precise orbits and relatively short temporal and spatial base. The scheme of processing is presented in Fig. 5.

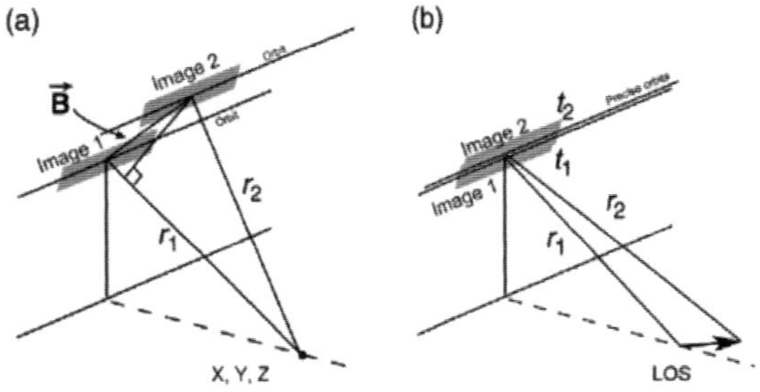

Fig. 4. Method of (a) radargrammetry INSAR DEM generation, and (b) DInSAR LOS displacement detection. Compilation based on [17].

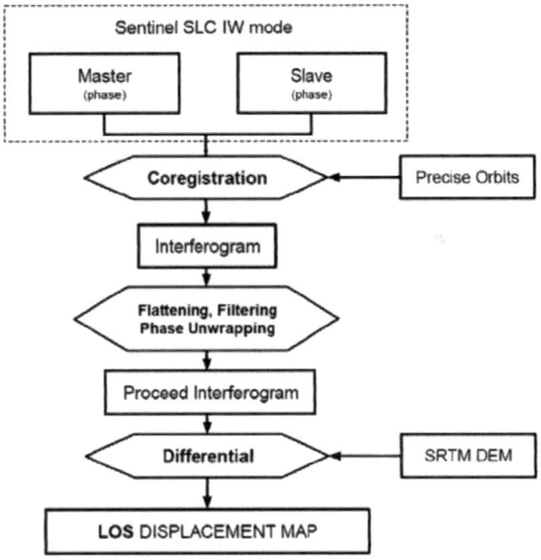

Fig. 5. Methodology of implemented DInSAR approach. Original compilation based on [18].

The data processing was performed in the GMTSAR software package [19], and the results are visualized with QGIS software [20]. The procedure of phase unwrapping was made with application of Snaphu algorithms [21]. Shuttle Radar Topography Mission (SRTM 1) data were used to correct of wave phase with regard to terrain surface [22].

5 Results

Since analyzed object is an open pit mine, the obtained map of terrain subsidence is based on a large number of points located in strong coherent regions. The SAR signal reflects well from an area uncovered by vegetation, both between 3–15.10.2016 and 15–27.10.2016. The LOS displacement map obtained after DInSAR data processing in the first analyzed internal is presented in Fig. 5. The processing indicates that the maximal subsidence values determined by the DInSAR method are -42 mm and occur in area C. This value is marked in green in Fig. 6. The Szczercow dumping site is an area where soil compaction takes place. The C-C' cross-section was generated following the analyses. It shows that the maximal value of terrain subsidence occurs in the top of the dumping site. Regions with potential terrain uplift during the 12 days of the analysis are marked in red. Terrain uplift occurs in excavation B and partly in excavation A, which may indicate relaxation following lignite exploitation. Subsidence occurs also in area A, in the eastern part of the excavation pit, which probably results from the occurrence of an internal dumping site and compaction of the deposited material.

The second analyzed period was based on imagery acquired between 15.10.2016-27.10.2016. The results of the compilation as Line of Sight (LOS) displacement maps are presented in Fig. 7. The maximal terrain subsidence for the analyzed period was -25 mm near the Szczercow dumping site. Moreover, the C-C' cross-section was generated for additional visualization. The maximal value of terrain subsidence occurs also in the top of the dumping site. The fact that soil compaction occurs also in the second study period indicates that the process may be continuous.

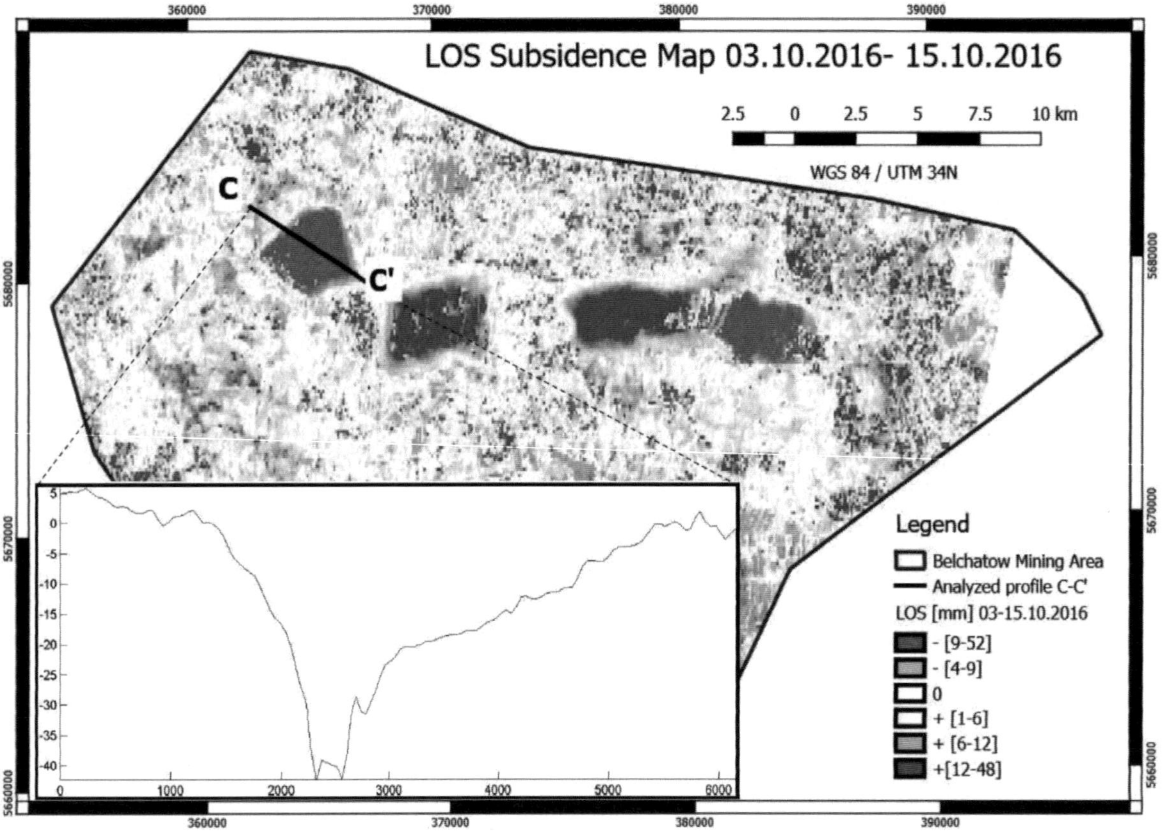

Fig. 6. Visualization of LOS displacements in the UTM system in zone 34N (EPSG: 32634) for a pair of SAR imageries acquired between 03.10.2016–15.10.2016, determined by the DInSAR method.

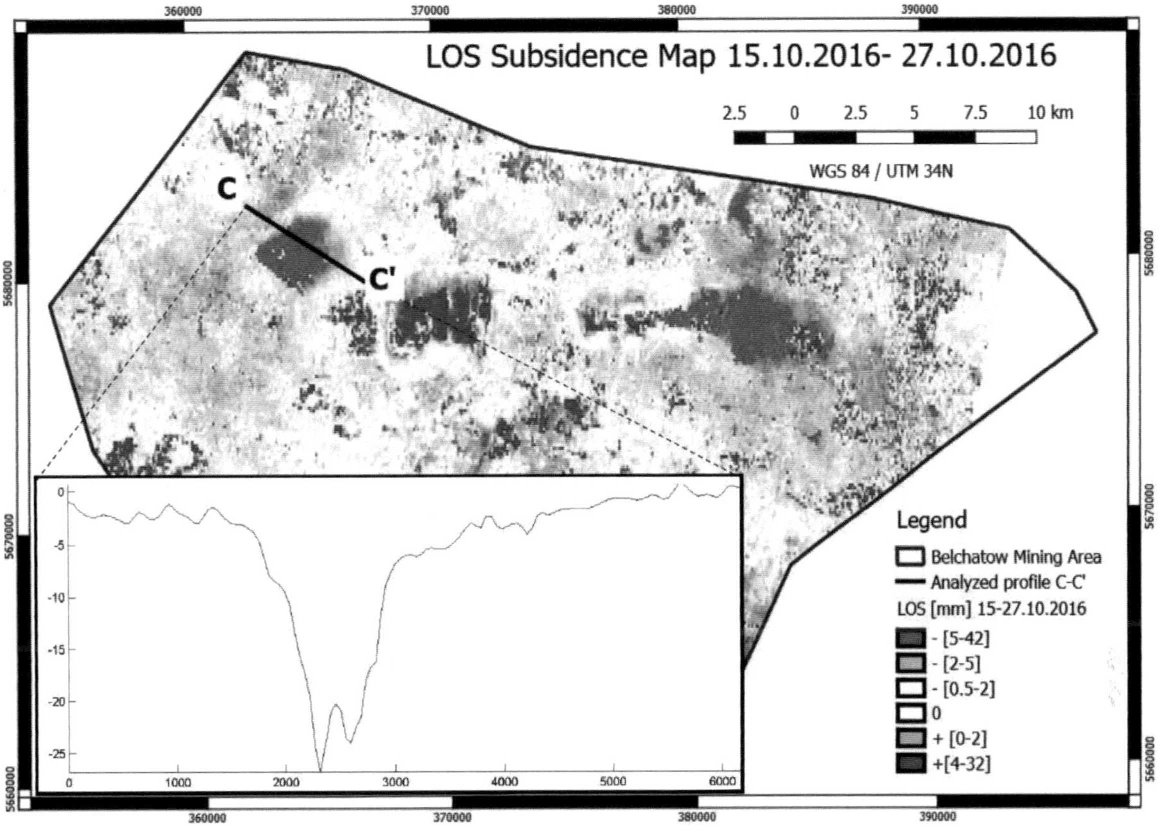

Fig. 7. Visualization of LOS displacements in the UTM system in zone 34N (EPSG: 32634) for a pair of SAR imageries acquired between 15.10.2016–27.10.2016, determined by the DInSAR method.

6 Summary

The paper presents the first experimental tests applying SAR imageries for the Belchatow open pit lignite mine. The research was focused on remote sensing sensor selection and elaborating a methodology allowing for correct data processing in a GMTSAR environment for the study object. The chosen Sentinel 1A SAR platform characterized by high temporal resolution was used in the study; in effect highly coherent interferograms were obtained. Areas with a similar characteristics of LOS displacement were registered for the analyzed two periods. These data represent a complete continuous surface information, as opposed to classical point surveys for the analyzed areas. Results of LOS displacement indicates that the Szczercow dumping site marked as C undergoes compaction in its top. Relaxation of the exposed lignite deposit in the Belchatow and Szczercow (A and B) excavation was also observed. Presented methodology allowed to detect continuous deformations with were presented at cross sections C-C'. The results show that the observed subsidence is relevant and allows to detect potential hazards. Further research will focus on processing of Sentinel 1A and 2A imageries, which should allow for obtaining a high-temporal resolution. It allows to acquire data every 6 days. More frequent sampling of the Belchatow open cast mine should allow more reliable results of the surface changes in the mining area to be observed. The next step should include data processing with application of time series INSAR analysis and validation of the results based on controlling classical land surveys for accuracy assessment.

This work was financed by the Polish Statutory Research Grant no. 0402/0123/17, 0402/0146/16 and 0402/0011/17.

References

1. J. Wajs, Mining Science, vol. **22**, 75-83 (2015)
2. C. Toth and G. Jóźków, ISPRS Journal of Photogrammetry and Remote Sensing, **115**, 22-36 (2016)
3. H. Peter-Contesse, A. Jäggi, J.J. Fernández, D. Escobar, F. Ayuga, D. Arnold, P. Visser, Advances in Space Research, Volume **60**, Issue 5, 879-892 (2015)
4. D. Massonnet and K.L. Feigl, Reviews of geophysics, **36(4)**, 441-500 (1998)
5. C. Carnec, D. Massonnet, C. King, Geophysical Research Letters, **23**(24), 3579–3582 (1996)
6. C. Colesanti, S. Le Mouelic, M. Bennani, D. Raucoules, C. Carnec, A. Ferretti. International Journal of Remote Sensing **26:1**, 201-207 (2005)

7. M. Costantini, F. Minati, M.G. Ciminelli, A. Ferretti, S. Costabile, Proceedings of IGARSS 2015, IEEE International Geoscience and Remote Sensing Symposium, Milan, Italy (IEEE Geoscience and Remote Sensing Society, 2015)
8. W. Milczarek, J. Blachowski, P. Grzempowski, Acta Geodyn. Geomater., Vol. **14**, No. 1 (185), 41-52 (2017)
9. S. Samsonov, N. d'Oreye, B. Smets, International Journal of Applied Earth Observation and Geoinformation, **23**, 142-154 (2013)
10. M. Crosetto, M. Castillo, R. Arbiol, Photogrammetric Engineering and remote sensing, **69(7)**, 775-783 (2003).
11. M. Mroz and Z. Perski, Proceedings of the 22nd Symposium of the European Association of Remote Sensing Laboratories, Prague, Czech Republic, 4-6 June 2002, 621-624 (2003)
12. M. Przyłucka, G. Herrera, M. Graniczny, D. Colombo, M. Béjar-Pizarro, Remote Sensing, **7(5)**, 5300-5328 (2015).
13. V.D. Poenaru, I.F. Dana Negula, A. Badea, R. Cuculici, The International Archives of the Photogrammetry, Remote Sensing and Spatial Information Sciences, Volume **XLI-B8**, XXIII ISPRS Congress, 12–19 July 2016, Prague, Czech Republic (2016)
14. E. Papadaki A. Tripolitsiotis, Ch. Steiakakis, Z. Agioutantis, M. Stelios, P. Partsinevelos, P. Schilizzi, Proc. First International Conference on Remote Sensing and Geoinformation of the Environment, Vol. 8795, 87951A (2013)
15. D. Colombo and B. MacDonald, Proceedings of the SAIMM International Symposium on Slope Stability in Open Pit Mining and Civil Engineering, Cape Town, South Africa (2015)
16. C. Ma, X. Cheng, Y. Yang, X. Zhang, Z. Guo, Y. Zou, Remote Sensing, **8(11)**, 951 (2016)
17. P.A. Rosen, S. Hensley, I.R. Joughin, F.K. Li, S.N. Madsen, E. Rodriguez, R.M. Goldstein, Proceedings of the IEEE, 88(3), 333-382 (2000).
18. A. Ferretti, C. Monti-Guarnieri, F. Prati, *InSAR Principles: Guidelines for SAR Interferometry Processing and Interpretation* (ESA Publications, 2007)
19. D.T. Sandwell, R.J. Mellors, X. Tong, M. Wei, P. Wessel, AGU Fall Meeting Abstracts (2010)
20. QGIS Development Team. QGIS Geographic Information System. Open Source Geospatial Foundation Project, http://qgis.osgeo.org (2018)
21. C.W. Chen and H.A. Zebker, J. Opt. Soc. Am. A, **17**, 401-414 (2000)
22. J.A. Slater, G. Garvey, C. Johnston, J. Haase, B. Heady, G. Kroenung, J. Little, Photogrammetric Engineering and Remote Sensing **72(3)**, 237-247 (2006)

Analysis of tectonic activity of the North Anatolian Fault based on SBInSAR method

Karolina Owczarz[1,*]

[1]Wrocław University of Science and Technology, Faculty of Geoengineering, Mining and Geology, 27 Wyb. Wyspiańskiego St., 50-370 Wrocław, Poland

Abstract. The North Anatolian Fault situated in Turkey is one of the longest and most active tectonic faults in the world. The paper presents an analysis of tectonic activity in its area based on the method of Small Baseline Synthetic Aperture Radar Interferometry. For this purpose 73 satellite SAR images and specialized software GMT5SAR were used with implement the SBAS algorithm. In addition, the most important aspects of data processing and their final products were presented, which determined the surface displacements occurring in the surveyed area from 1 January 2014 to 1 March 2017. The displacements of the SBAS surface area ranged from -10 cm to +10 cm. Based on the obtained results and their analysis, the author also assessed the suitability of SBInSAR technology for areas of land displacement.

1 Introduction

The tectonic activity of the terrain is related to the geological structure of the Earth, as well as to the endogenous processes, which have great influence on the topography. The theory of lithosphere tectonics developed in the early 1960s by Harry Hammond Hess assumes that the lithosphere consists of smaller and larger plates forming the continental crust and the oceanic crust. Under the lithosphere, there is an asthenosphere or layer of the upper part of the Earth's mantle, characterized by high plasticity and reduced velocity of seismic wave propagation. The plastic properties of the asthenosphere allow the movement of lithosphere plates and isostatic movements. In the asthenosphere, there are convection currents responsible for the wandering of magma in the interior of the Earth. Lithospheric plates move and are accompanied by pressure or lateral stresses, which cause the formation of new, tectonic forms of the relief, such as: mountains, volcanoes, faults, oceanic trenches, etc. The geomorphology research new landforms, determines their shape, geometric features, provenance, mechanism of development and age. To determine the tectonic activity within terrain forms measurements: geodetic (leveling, tachymetry, laser scanning, GNSS), seismic, geophysical or remote sensing are made.

Technique Small Baseline Synthetic Aperture Radar Interferometry (SBInSAR) also called Small Baseline Subset (SBAS) is one of many remote sensing methods the satellite radar interferometry. SBInSAR method examines small, to long-term land displacements, for example: tectonic faults, slides, volcanoes, water reservoirs, glaciers, building objects, subsidences or uplifts terrain, mining and post mining area, etc. SBAS is derivative of Differentiall Synthetic Aperture Radar Interferometry and allow to detection slow movements at the milimeter level by difference between the main interferogram and the synthesized interferogram. Currently SBAS makes to the latest algorithm used to monitoring surface displacement, which generates multiple pairs of interferograms with small spatial and temporal baseline. This minimizes the effects spatial and temporal decorrelation and topographic artifacts due to the use of Digital Elevation Model (DEM). A detailed description of the SBInSAR method was presented in publications [1, 2].

The use of satellite images to determine the topography of the Earth and the physical processes that occur on it are of great interest, as evidenced by numerous scientific publications on this subject [3, 4]. A study of the interactions between permafrost and infrastructure along the Qinghai-Tibet Railway was presented in article [5]. Based over a dozen images and application of the SBAS method, they analyzed the relationship between surface displacement, permafrost active layer and environmental conditions. In this project, the researchers estimated that surface motions along embankment are in the range of 20mm/yr to +20mm/yr. The issue of monitoring land subsidence and fault displacement occurring on the Datong Basin (China) has appeared in the publication [6]. For the SBInSAR calculation 40 radar scenes were used and were taken the relationships among the regional land subsidence, ground fissures and faults activity: Kouquan, Liulengshan Premont and Hengshan Premont. As a result of the displacement studies conducted in the Datong Basin, the variation range -15.5mm/yr to +10.4mm/yr. The SBAS method was also used to detect and monitor subsidence of land in the Nordic countries [7]. During

* Corresponding author: karolina.owczarz@pwr.edu.pl

the research, the results of settlements at the level of 5 mm/year were used. It turns out that SBInSAR is not only an appropriate tool for detecting displacements, but also for assessing urban policy and supporting the design and implementation of effective development strategies [8].

This article focuses on the determining surface displacement in the North Anatolian Fault Zone (NAFZ) from 1 January 2014 to 1 March 2017 by acquiring satellite images, using SBInSAR technique and developing results.

2 Study area – North Anatolian Fault

The North Anatolian Fault (NAF) is located in the northern part of Turkey and extends from the east to the west of the country (Fig. 1). This is one of the longest tectonic faults in the world. It was first described in the late 1940s by I. Ketin [9] and is one of the most well studied zones of tectonic activity [10-12]. NAF is a particularly dangerous geological formation that has caused massive damage in the Marmara Sea region.

There are numerous earthquakes documented over the centuries, and the 1999 Izmite disaster has led geologists to discover that the North Anatolian Fault are numerous smaller faults that create a dynamically connected tectonic form of a single 1,600 km long fault. Thus, whenever there are strong tremors in the region, their energy is transferred along the rock fissures to other areas [14].

NAFZ is an intra-continental, strike-slip, transform fault that forms the geological boundary between the Anatolian plate and Eurasian plate. Its right-lateral slip is clearly due to physiographic conditions and geological features. The fault creates a narrow fissure running parallel to the Black Sea coast, starting from the Karliova triple junction in the east, to the town of Geyve in the west, where it divides into three sections in the eastern part of the Marmara Sea. These three bands stretch west to the bays: Saros and Edremit, with the northern section running through the Gulf of Izmit [10].

The origin of the NAF is controversial, although the most probable theory is that it originated as a result of the partial closure of the southern part of the Tethys Ocean along sector between present-day Bitlis town in eastern Turkey and the Zagros mountains extending east of Iran during the middle Miocene. The fault formed earlier in the zone of closure of the northern part of the prehistoric ocean between Paleocene and Eocene. The structure of the NAF was formed shortly after the continental collision between the Arabic plate and the Eurasian in eastern Anatolia [11]. Due to the above events, the fault has so far been characterized by a varied geological structure on both sides of the formation.

Due to the length of the NAF was exposed only the fragment between 32° and 33° east longitude (Fig. 2). According to the division of NAF proposed in the publication [12] this fragment belongs to the central part of the structure in the area, where in the XX century there were several earthquakes of magnitude over 7.

Therefore, the project was limited to only fragment of the area along the NAF, covered by swath no. 3, to verify the suitability of the SBAS method for analyzing tectonic activity and to obtain the values of displacement occurring in the area under investigation.

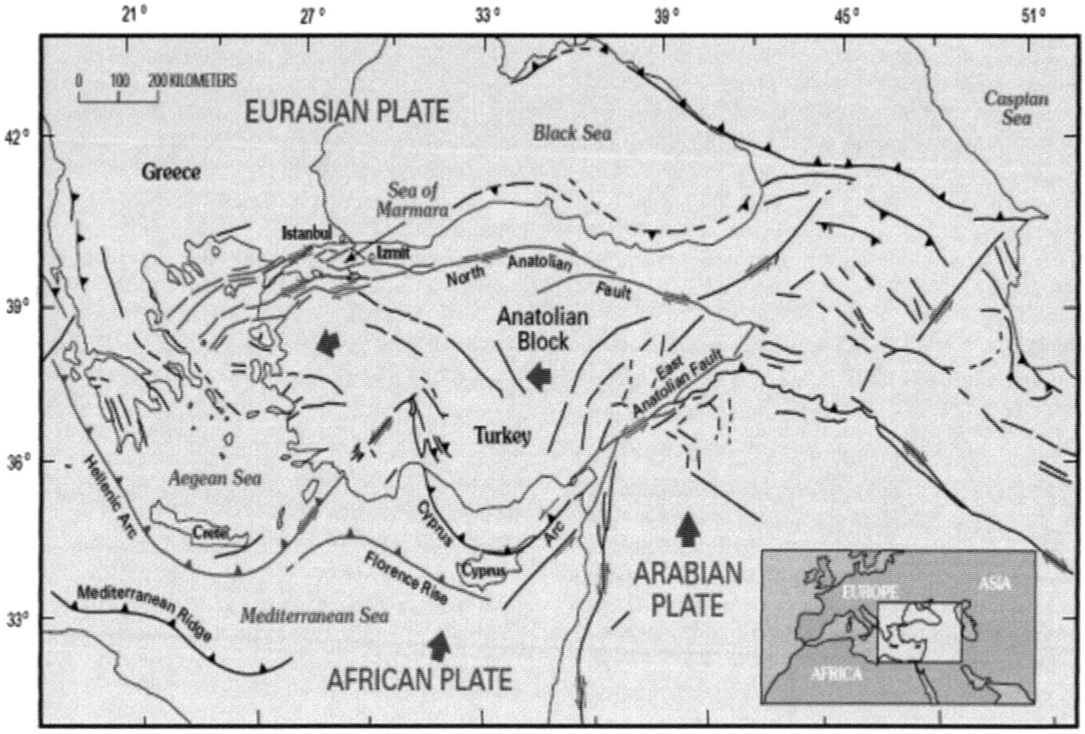

Fig. 1. Geological map of Turkey [13].

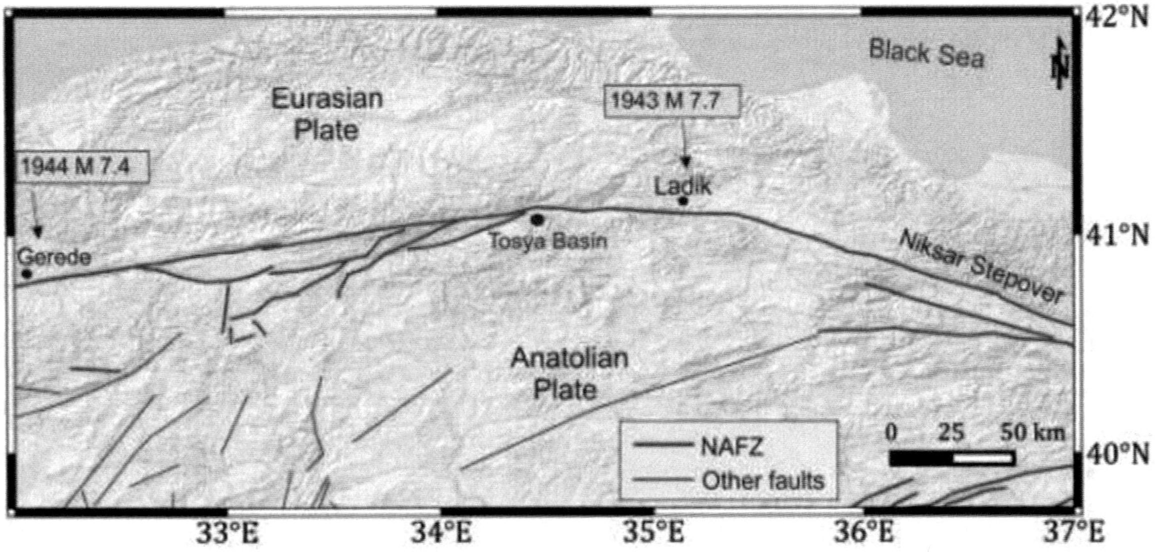

Fig. 2. Map of the central part of the NAFZ [12].

3 SAR data collections of the NAF

The project uses satellite data registered by the Sentinel-1A and Sentinel-1B satellites, assume of the Copernicus program under the European Space Agency's in the period from 1 January 2014 to 1 March 2017. In addition to satellite images, in SBAS technique was required to obtain a Digital Elevation Model and precise satellite ephemeris. Radar images consisted of 73 scenes (Figs. 3, 4A) including 60 from Sentinel-1A and 13 from Sentinel-1B and were characterized by following parameters: ascending/descending, Single Look Complex (SLC), Relative Orbit Number (167) and VV polarization.

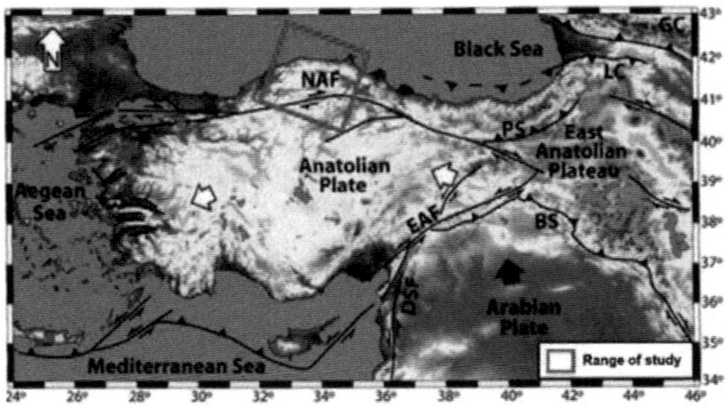

Fig. 3. Range of downloaded satellite images [15].

Also was used the DTM (Fig. 4B) of the Shuttle Radar Topography Mission (SRTM) with a resolution of approximately 1" (30 meters) as a script. The Digital Elevation Model was reduced to a synthesized interferogram eliminating the components responsible for topography [16]. The last group of necessary data was information on the precise position of the satellites, because in particular in the case of interferometry, the Sentinel-1 requires meticulous control.

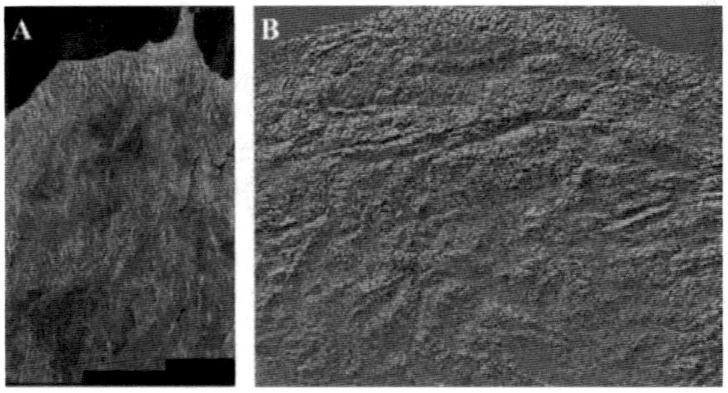

Fig. 4. A) The satellite image [17], B) Digital Elevation Model [18].

The above data came from the following web portals: *Copernicus Open Access Hub* – satellite images, *GMTSAR* – DTM as a script and *Sentinel-1 PDGS* – precise position of the satellites providing complete, free and open user access to the products. The most important element was the proper preparation of data by organizing them in such a way that they could be processed in batch files.

4 SBAS data processing

The SBInSAR is a DInSAR algorithm for detecting displacement of the Earth's surface and analyzing its time evolution by generating average velocity maps and time series of displacements in direction line of sight - LOS [19]. In particular, the method consists in the creation of interferograms by the appropriate choice of SAR data pairs characterized by small time thresholds and small spatial baselines [2]. This is to mitigate the effects of signal decorrelation, and thus maximize the number of coherent pixels in terms of time. SBAS allows limiting the impact of topographical artifacts and includes filtering elements of atmospheric phases based on the observation of their signal, which is strongly correlated in outer space but less in time [1].

The SBAS method was described in detail in publication of [1], in which the entire calculation scheme is presented mathematically. The following is a brief description of the individual implementation processes of the Casu [20] SBAS algorithm:

Stage 1: Evaluation of the orbital parameters associated to each SAR acquisition (subsequently used within the estimation of the spatial baseline value) and reference Master image selection.

Stage 2: Generation of a set of single look SAR images (SLC) from the available raw data files (focusing).

Stage 3: DEM conversion into reference Master image SAR coordinates and computation of sensor to target distance files (range files) for each acquisition.

Stage 4: Optimal interferometric data pair distribution selection.

Stage 5: Co-registration of each SLC data pair with respect to the selected reference Master image.

Stage 6: Differential interferograms and corresponding spatial coherence maps generation.

Stage 7: Noise-filtering of the generated DInSAR fringes.

Stage 8: Generation of the mean displacement velocity map and the corresponding time series, for each coherent pixel of the investigated area, via the inversion of the computed sequence of DInSAR interferograms. At the same time, an estimate of the possible residual topographic components and the atmospheric contributions is also accomplished;

Stage 9: Geocoding of the obtained results and projection into a universal cartographic grid.

All calculations were done in the GMT5SAR software, where was made the so-called batch processing that is the automatic processing of large SAR data volumes to generate interferograms (Fig. 5).

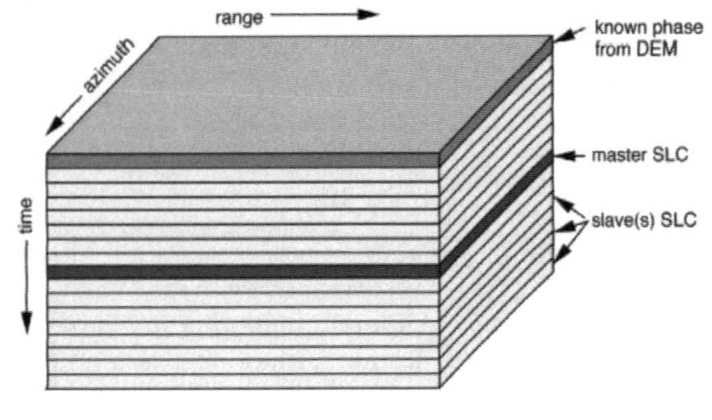

Fig. 5. Batch processing for time series [21].

The entire processing was implemented in accordance with the individual stages highlighted in the SBAS method and with appropriately selected input parameters for the studied area. Due to the large number of satellite imaging, more than 400 interferograms representing the phase difference between two radar scenes were obtained. The most important stage was the development of the phase consisting in completing the measured phase by a multiple of 2 π, to then proceed with the calculation of cumulative displacements.

5 Analysis of results

Turkey is a seismically active area as evidenced by the scientific articles [22, 23]. This activity is due to the fact that Turkey is located at the junction of several lithosphere plates (Eurasian plate, African plate, Arabic plate and Iranian plate). The central part of the fault compared to the eastern and western parts exhibits much less seismic shocks [12]. As a result of the SBAS algorithm, 73 maps cumulative displacements between successive satellite data acquisition were obtained. The article was presented selected ones at several monthly intervals between the first and the last image (Fig. 6). The maps below do not have a fixed time interval. Only maps that show significant displacement are selected.

On this basis, the tectonic activity of the examined area was analyzed. In addition, four points A, B, C and D were applied to check their individual displacements. Points A and B were located at the beginning and end of the analyzed fault segment. Points C and D were placed respectively north and south of the fault zone to see if the changes occurring in the fault strongly affect the surrounding area. The following displacement maps refer to two data groups (three maps for each set) created during the calculation.

By analyzing the cumulative displacements comparison, it is clear that the variations in the surveyed area range from -10 cm/yr to +10 cm/yr, with SBAS accuracy being able to reach millimeters [24]. The dominant part is the uplifts (green and yellow hues), while there are few settlements of the surface (blue hues). Displacements occur at the points of the fault line. Below is a chart of cumulative time displacements for four selected points (Fig. 7): A, B, C and D, which were analyzed. The points were placed in such space of the study area that it was possible to determine the variability of the changes. Thus, the largest displacements from -1.4 cm to +2.3 cm show the B point located in the eastern part of the area, where the branching is visible on several new fault lines. Likewise, the situation for point C is placed in the north, whose values oscillate range from -1.5 cm to +1.5 cm. Point A, despite being located in the western fragment of the fault line, is characterized by quite small displacements – from the level of +0.4 cm to +1.5 cm. On the other hand, point D of all shows the smallest changes during the entire observation period, since they fall within the range of -0.2 cm to +0.6 cm. Analyzing the points in pairs A-D and B-C results the dependence characterizing the course of displacements in the examined period. Namely, the displacements of points A and D tend to increase. However, this point A reaches higher values. In the case of the B-C graph, there are significant changes in time, as the displacements from the smallest negative values pass into the highest positive values. This state occurs up to the half of the analyzed period, while later the displacement gradually increases. Point B is characterized by greater displacements from point C.

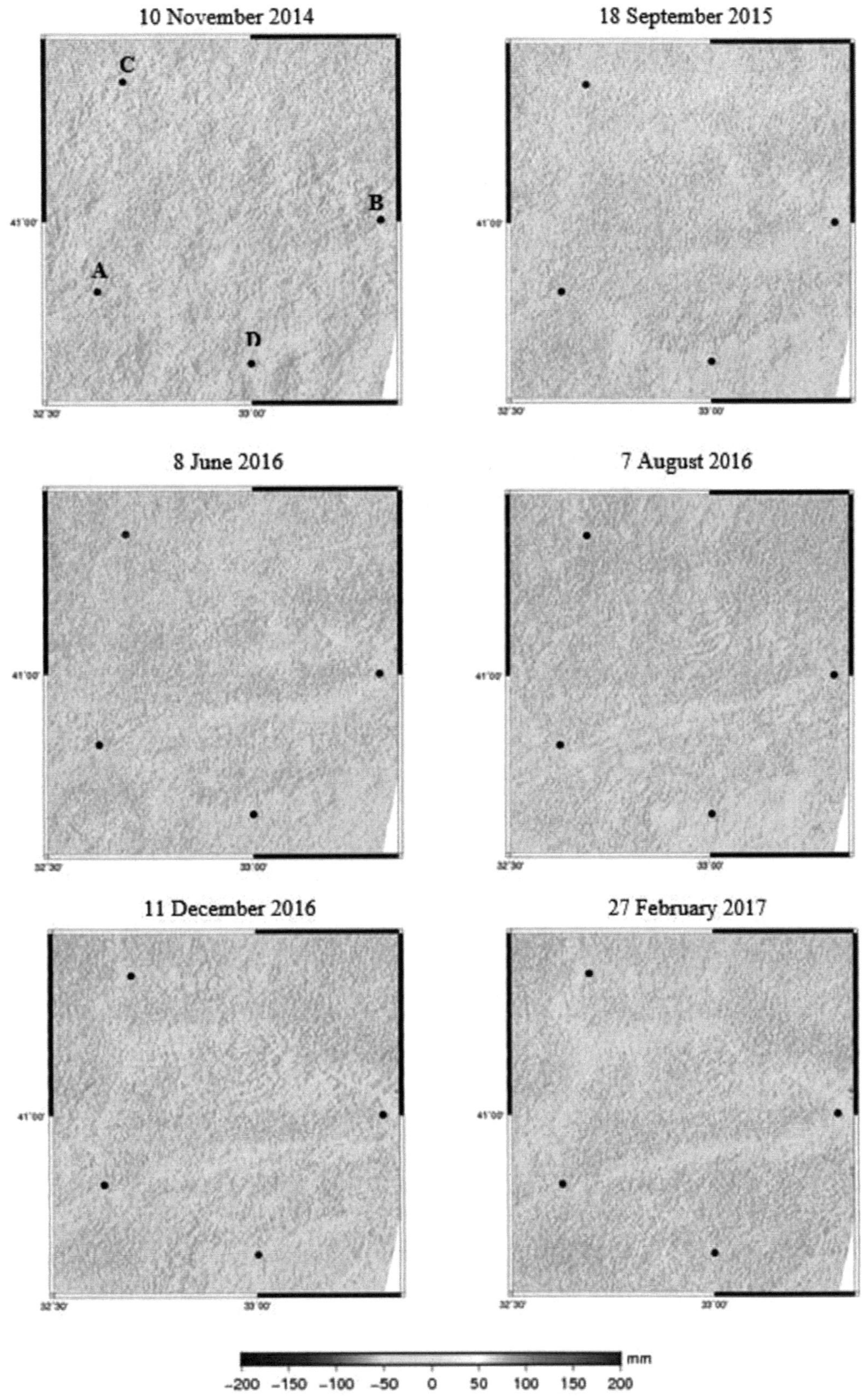

Fig. 6. Cumulative displacement in region the NAFZ from November 2014 to February 2017.

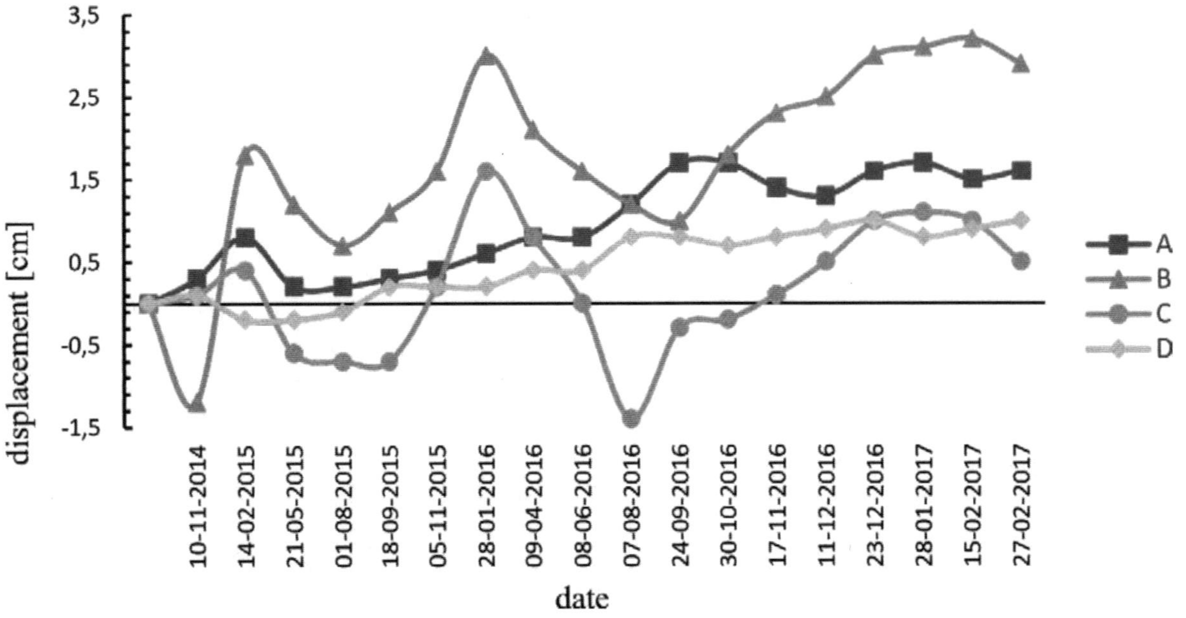

Fig. 7. Cumulative displacement in region the NAFZ or points: A, B, C and D from November 2014 to February 2017.

Recapping the results of the analysis, the northern part of the area compared to the southern part shows greater displacements. Visible branches on the generated maps are related to the fact that in Turkey there is a system of many faults dynamically connected together with NAF. In this way, the stresses that are released in the earth's crust are also able to spread more rapidly to vast areas. In the case of Turkey it should be emphasized that most of the manifestations of tectonic activity are due to collisions of lithosphere plates. Its part located south of the North Anatolian Fault and west of the East Anatolian Fault moves to west. Unlike some parts of Turkey outside of that zone, which is located on the Anatolian continental plate and moves west. This movement causes the pressure of the African plate that moves north. In addition, the Arabic plate also causes complications in contact with the Anatolian and Eurasian plates, creating a folding – faulting the Bitlis – Zagros zone in the eastern part of the country.

For in-depth analysis, it would be useful to use data from the North Anatolian Fault along the entire length of the radar interferometry. This would allow for a better understanding of the changes that are taking place in this area of the world.

6 Conclusions

This article presented the application of Synthetic Aperture Radar Interferometry SBAS to determine the displacements of a selected fragment of the terrain in the area of the North Anatolian Fault in Turkey. Measurements were made using specialized software GMT5SAR. Selection of the SBAS method was due to the fact that this technique is a characterized by the minimized influence of atmospheric artefacts and possible topographic errors. In addition, it allowed selections of both baseline and temporal distances, on the basis of which interferograms were generated between pairs of images. By defining these two parameters it was possible to eliminate spatial and temporal decorrelations. As a result of data processing, the SBInSAR method has been shown to be significantly effective in detecting displacements of the selected fault zone over a longer period of time – over 2 years.

In order to determine the displacements of the North Anatolian Fault, 73 satellite images of Sentinel 1 were used, which were processed using the SBAS method. Based on the obtained results, the tectonic activity of the studied area was analyzed. The cumulative displacements in the center of the fault range from -10 cm to 10 cm. In addition, the movements of several points located in this area have been analyzed, whose displacements oscillate from -1.5 cm to 3.5 cm. The above results indicate the occurrence of the phenomenon of uplift and subsidence of the surface. The variations in altitude of the surface can be related to the geological processes responsible for significant seismic activity, referred to as the collision of lithosphere plates. It is precisely the continental plates that, through their movements, in the Earth's crust, cause considerable tension, often leading to earthquakes. The North Anatolian Fault as a geological boundary between the Anatolian plate and the Eurasian plate continually undergoes their influence, leading to its displacements.

Based on the conducted work, it has been proved that it is possible to obtain high accuracy of results and information on changes in topography of the area for the studied area by applying the SBInSAR method. This technology is characterized by the availability of both input materials and the required software, but requires good hardware with high computing power. Final products can be processed in any way according to user requirements. Their digital form provides the opportunity to return at any time to the various stages of work. The SBAS method allows detecting displacement of various types of land (faults, volcanoes, slides) and the accuracy of results can be increased by combining it with other methods such as GPS or PSInSAR.

References

1. P. Berardino, G. Fornaro, R. Lanari, E. Sansosti, IEEE Transactions on Geoscience and Remote Sensing, **40**, 2375 - 2383 (2002)
2. R. Lanari, F. Casu, M. Manzo, G. Zeni, P. Berardino, M. Manunta, A. Pepe, Pure and Applied Geophysics, **164**, 637-661 (2007)
3. P. Shanker, F. Casu, H.A. Zebker, R. Lanari, IEEE Transactions on Geoscience and Remote Sensing, **8**, 592-596 (2011)
4. C. Zhao, Q. Zhang, C.Yang, J. Zhang, W. Zhu, F. Qu, Y. Liu, International Association of Hydrological Sciences, **372**, 305-309 (2015)
5. F. Chen, H. Lin, Z. Li, Q. Chen, J. Zhou, Remote Sensing of Environment, **123**, 532-540 (2012)
6. Y. Cheng-sheng, Z. Qin, Z. Chao-ying, W. Qing-liang, J. Ling-yun, Journal of Geodynamics, **75**, 34-40 (2014)
7. T.R. Laukens, J.Dehls, Y.Larsen, K.A. Høgda, D.J. Weydahl, Fringe Workshop (2006)
8. F. Calò, S. Abdikan, S. Görüm, A. Pepe, H. Kiliç, F.B. Şanli, Remote Sensing, **7**, 16519-16536 (2015)
9. I. Ketin, Geologische Rundschau, **36**, 77-83 (1948)
10. Z. Cakir, S. Ergintav, A.M. Akoğlu, R. Çakmak, O. Tatar, M. Meghraoui, Journal of Geophysical Research: Solid Earth, **119**, 7934-7943 (2014)
11. A.E. Aksu, T.J. Calon, R.N. Hiscott, GSA Today (2000)
12. M. Bohnhoff, P. Matrinez-Garzón, F. Bulut, E. Stierle, Y. Ben-Zion, Tectonophysics, **674**, 147-65 (2016)
13. J.L. Gross, L.T. Phan, U.S. Geological Survey, **1193**, 64 (2000)
14. R. Gore, National Geographic Polska, (2015)
15. A.A. Özacar, G. Zandt, H. Gilbert, S.L. Beck, Geological Society, **340**, 485-496 (2010)
16. B. Rabus, M. Eineder, A. Roth, R. Bamler, ISPRS Journal of Photogrammetry & Remote Sensing, **57**, 241-262 (2003)
17. https://scihub.copernicus.eu/dhus/#/home (2017)
18. http://topex.ucsd.edu/gmtsar/demgen/ (2017)
19. M. Manzo, Y. Fialko, F. Casu, A. Pepe, R. Lanari, Pure and Applied Geophysics, 169, 1463-1482 (2011)
20. F. Casu, *The Small BAseline Subset technique: performance assessment and new developments for surface deformation analysis of very extended areas* (Dept. of Electrical and Electronic Engineering University of Cagliari, 2009)
21. T. Xiaopeng, D. Sandwell, *InSAR workshop*, Conference materials (2013)
22. F. Bulut, M. Bohnhoff, W.L. Ellsworth, M. Aktar, G. Dresen, USGS Staff - Published Research, **114**, 16 (2009)
23. M. Bohnhoff, F. Bulut, G. Dresen, P.E. Malin, T. Eken, M. Aktar, Nature Communications, **4**, 1999 (2013)
24. P. Berardino, F. Casu, G. Fornaro, R. Lanari, M. Manunta, M. Manzo, A. Pepe, E. Sansosti, *Small Baseline DIFSAR Techniques for Earth Surface Deformation Analysis* (2003)

LOS Displacements of Mauna Loa volcano, Hawaii Island, as determined using SBAS-InSAR

Karolina Grzesiak[1], and *Wojciech J.* Milczarek[1,*]

[1]Wrocław University of Science and Technology, Faculty of Geoengineering, Mining and Geology, 27 Wyb. Wyspiańskiego St., 50-370 Wrocław, Poland

Abstract. This paper presents an overview of the surface displacements retrieval capability of the Differential Synthetic Aperture Radar Interferometry algorithm, using Small Baseline Subset (SBAS) technique, in the context of active volcanic areas. Based on Sentinel-1A images, time series displacements fields of the Mauna Loa volcano area over a 2-year time interval (between 2015 and 2017) were received using batch processing. Based on 35 radar images, a total of 179 interferograms have been calculated. Methodology of Synthetic Aperture Radar (SAR) Sentinel-1 ESA satellite mission data processing with small base (SBAS) interferometric techniques from has been presented. The displacements calculated in the satellite's Line of Sight (LOS) have been presented graphically on maps and graphs. Application of radar interferometry methods in the case of volcanic surface activity research creates new possibilities in the area of permanent monitoring of this type of areas.

1 Introduction

The present shape and location of the continents is the result of tectonic plates movements over millions of years. Numerous tectonic processes that lead to morphological differences in the terrain are observed at the boundaries of the plates. One of them is volcanism, which can both create new forms and - in the case of extremely explosions - lead to the destruction of previous forms. The earthquake is a frequent phenomenon preceding the eruption of the volcano [1]. Advanced measurement techniques enable us to obtain information about the elevation shape, observe a lot of phenomena and examine the physical properties of the Earth surface. Data for this type of research is acquired direct or indirect methods. The first group includes data from direct measurements such as GNSS, levelling, tachymetric and terrestrial laser scanning, while the second group includes radar images acquired by satellites. The range of the aforementioned methods largely depends on the quality and technical requirements, the availability of data, the physical characteristics of the object under study, and time. Remote sensing is useful in monitoring difficult to reach areas where stationary testing is dangerous, impossible to execute, or too costly. Radar interferometry is used to obtain information about the topography of the study area, as well as an excellent tool for monitoring displacements of terrain. These changes are mainly due to natural phenomena such as earthquakes, volcanic activity, landslides and human activities.

This article was created during the research work within the thesis. The main objective was to investigate Mauna Loa's volcanic activity over two years, adequate for the operation of the Sentinel – 1A satellite, to check the suitability and quality of results based on sentinel radar images. For this purpose, 35 satellite images were processed using the open source GMT5SAR [2] program, available only on the UNIX operating systems.

2 Mauna Loa volcano – basic information

Mauna Loa is a shield volcano located on the southwestern side of the Hawaiian Archipelago Island – Hawaii and is one of the largest active volcano in the Earth (Fig. 1). Along with the other volcanoes, it is under constant observation by the USGS Research and Development Agency. Mauna Loa occupies most of the island's total area (about 5180 km^2), while it measures 4168 m above the Pacific Ocean surface [3]. The last eruption was in 1984 and it has erupted 39 times since 1832. Based on research and geological analyses, it has been shown that below the Pacific plate, there are still hot spots. The manifestation of this phenomenon allowed the formation of volcanic islands and volcanoes, by collecting the lava extracted from the interior of the land. Currently, the hot spots are located directly below the central part of the Hawai'i Islands, causing the activity in Mauna Lao and Kilauea volcanoes of the five volcanos. As a result of the eruption, the volcano produces low – viscosity basaltic lava (fluid and dark). The sudden drop in pressure and temperature contributes to the crystallization of the lava, the characteristic chemical composition of 'hawaiian' lava, which causes it to move over long distances, thus forming the outer crust of a

* Corresponding author: wojciech.milczarek@pwr.edu.pl

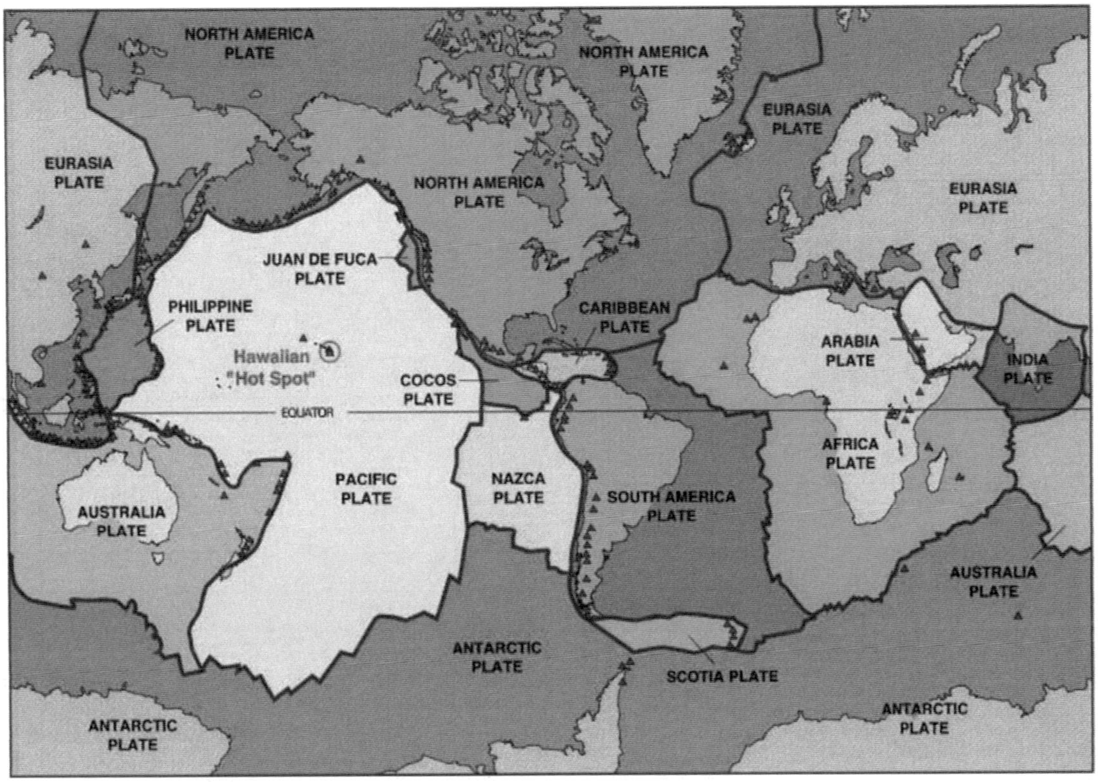

Fig. 1. The map of volcanic phenomena on global tectonics background [4].

volcano. Mauna Loa exhibits its activity more or less effectively from the dawn of history [4]. Based on numerous observations, three basic volcanic tidal zones have been distinguished for their range. The most frequent eruptions and lava flows occurred in Northeast Rift Zone and Southwest Rift Zone [5]. Various measuring techniques are used to observe volcanic activity, which can be used to the detect changes that precede volcanic eruptions. The monitoring of the volcanic area involves continuous or real – time observations, such as earthquakes, ground movements, volcanic gases. Observation is carried out using surveying instruments, GSP receivers, satellite images, cameras, thermal imaging cameras and seismic sensors. Collecting and interpreting information from different measurement techniques leads to an effective and consistent outcome for the changes in the object. Areas of the earthquake are mainly caused by eruptions within active volcanoes or deep structural changes due to the mass of islands on the earth's crust. Based on research, the HVO has shown that the vast majority of earthquakes occur in the southern part of the island, where Mauna Loa and Kalauea volcanoes are located. These quakes are divided into three classes resulting from [6]:

- the movement of magma inside volcanoes, eruptions on peaks or rift zones and the accumulation of magma in shallow reservoirs,
- plate movement within the tectonic faults,
- cambered or folded earth crust and upper mantle.

3 Data

Previously acquainted with the technique of generating and determining displacements, it was possible to select the appropriate input data. Data is divided into three groups: satellite images, DEM (Digital Elevation Model): the Shuttle Radar Topography Mission (SRTM) version 1 [7] and precise orbit ephemerides. Satellite images provided by the European Space Agency (ESA) were used to analyse the volcanic activity. High quality data is available on the website, which is updated on a regular basis. The data set consisted of 35 images (with perpendicular baselines of less than about 70 m) captured by Sentinel – 1A satellite, moving in ascending orbits number 124 (relative orbit) between May of 2015 and June of 2017 (Tab. 1). The interval between registrations was mostly 12 days, but in occasional cases it was necessary to skip images – images didn't represent the whole object. Due to the presence of vegetation cover (Forest Reserves) Sentinel – 1A data are used, which operated on the C band. Sentinel – 1 potentially operates in four exclusive acquisition modes: stripmap (SM), interferometric wide swath (IW), extra-wide swath (EW) and wave (WV). SAR instrument supports operation in single and dual polarisation. For ground displacement measurements, registration in IW mode is used – so each satellite images consists of three sub – swaths that have been processed as SLC products. IW mode acquires data with a 250 km swath at 5 m by 20 m spatial resolution. The analysed area is located in the central sub – swaths of IW2 (Fig. 2).

https://doi.org/10.1051/e3sconf/20185500006

Table 1. Sentinel – 1A SAR acquisitions on ascending orbits and images parameters.

Mission	Date of registration	Orbit	Resolution	Polarisation	Mode	Sub-swath
S1A	30/05/2015	6146	SLC	VV	IW	IW2
S1A	11/06/2015	6321	SLC	VV	IW	IW2
S1A	23/06/2015	6496	SLC	VV	IW	IW2
S1A	17/07/2015	6846	SLC	VV	IW	IW2
S1A	29/07/2015	7021	SLC	VV	IW	IW2
S1A	22/08/2015	7371	SLC	VV	IW	IW2
S1A	03/09/2015	7546	SLC	VV	IW	IW2
S1A	15/09/2015	7721	SLC	VV	IW	IW2
S1A	09/10/2015	8071	SLC	VV	IW	IW2
S1A	21/10/2015	8246	SLC	VV	IW	IW2
S1A	25/01/2016	9646	SLC	VV	IW	IW2
S1A	18/02/2016	9996	SLC	VV	IW	IW2
S1A	01/03/2016	10171	SLC	VV	IW	IW2
S1A	25/03/2016	10521	SLC	VV	IW	IW2
S1A	12/05/2016	11221	SLC	VV	IW	IW2
S1A	24/05/2016	11396	SLC	VV	IW	IW2
S1A	05/06/2016	11571	SLC	VV	IW	IW2
S1A	29/06/2016	11921	SLC	VV	IW	IW2
S1A	11/07/2016	12096	SLC	VV	IW	IW2
S1A	23/07/2016	12271	SLC	VV	IW	IW2
S1A	04/08/2016	12446	SLC	VV	IW	IW2
S1A	16/08/2016	12621	SLC	VV	IW	IW2
S1A	09/09/2016	12971	SLC	VV	IW	IW2
S1A	15/10/2016	13496	SLC	VV	IW	IW2
S1A	27/10/2016	13671	SLC	VV	IW	IW2
S1A	08/11/2016	13846	SLC	VV	IW	IW2
S1A	02/12/2016	14196	SLC	VV	IW	IW2
S1A	26/12/2016	14546	SLC	VV	IW	IW2
S1A	07/01/2017	14721	SLC	VV	IW	IW2
S1A	31/01/2017	15071	SLC	VV	IW	IW2
S1A	12/02/2017	15246	SLC	VV	IW	IW2
S1A	24/02/2017	15421	SLC	VV	IW	IW2
S1A	08/03/2017	15596	SLC	VV	IW	IW2
S1A	20/03/2017	15771	SLC	VV	IW	IW2
S1A	01/04/2017	15946	SLC	VV	IW	IW2

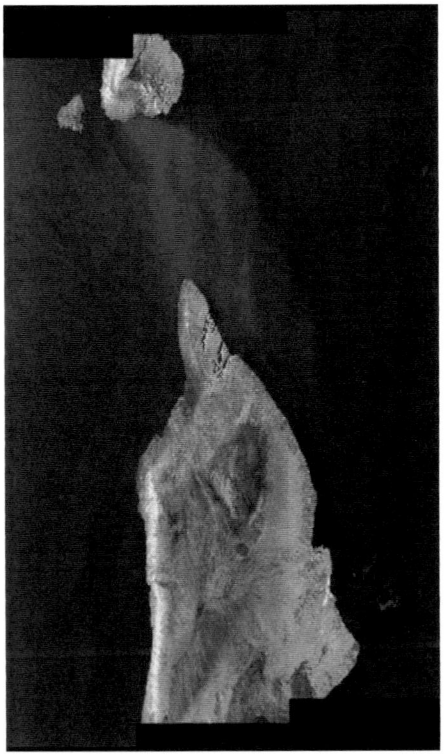

Fig. 2. Sentinel 1A image example from Hawaii Island with sub – swaths.

Another essential input file is the Digital Elevation Model – a digital model or a graphical representation of the topographic surface of the ground (Fig. 3). The model is characterized by a high resolution of 30 m and homogeneity in accuracy.

To generate satellite image products, the processor requires additional information not included in the satellite data. Therefore, it was necessary to retrieve auxiliary data from the PDGS – precise orbits ephemerides. The files should match the previously downloaded images for the time information.

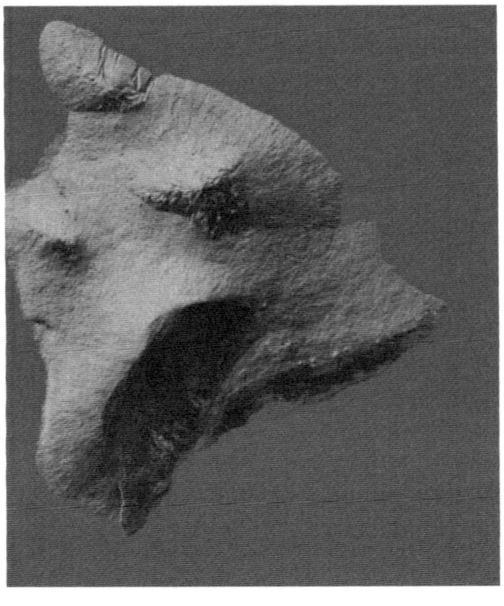

Fig. 3. Digital Elevation Models for Hawaii Island - SRTM 1.

4 InSAR – how does it work

InSAR method is used to measure changes in the morphology of the Earth's surface using satellite images. Information about ground changes in time interval, is included in images composed of interference bands. Based on two radar images, the phase difference for each pixel is determined, thus creating an interferogram. The selected radar images between the designated displacements should represent the same area but differ in the date registration. Determining the phase difference, attention should be paid to reflective properties of the surface, which in each image should assume equal value. Phase difference values can be obtained in single pass or repeat pass interferometry. The essential difference between modes is the number of necessary radar transits needed to obtain images. Repeat pass interferometry is a method of data registration on a single satellite only during two independent flights. Therefore, it is necessary to obtain accurate information about localization of the satellite during the registration. Based on them a spatial baseline is defined (perpendicular baseline and parallel baseline). Exceeding the critical value of the baseline makes it impossible to generate interferograms based on the selected images. Critical values depend on radar wavelength, beam angle, horizontal resolution and distance between radar and surface. The adjusted parameters determine the geometric dependence of the position of the satellites during the imaging process. Analysis of displacements and ground changes is based on the displacements phase determined, so the effect of other factors should be removed (for example topography, ground movements, atmospheric conditions, spatial distances between satellites. Elimination of the topography component can be achieved by using the DEM model. The model is converted into a synthesized interferogram. The source of interference in InSAR analyses is atmospheric artifacts and spatial decorrelation. The atmospheric artifacts are highly correlated in space but poorly in time. The deviation of the phase value can be caused by the heterogeneity of pressure, temperature or water in the troposphere. Tropospheric artifacts are present in repeat pass interferometry products, as atmospheric conditions change over time. By contrast, temporal decorrelation is related to the occurrence of vegetation, humidity changes and abundant rainfall or snowfall. It has a significant impact on the quality of interferograms generated from time baseline radar images.

5 Method of InSAR data processing – SBAS algorithm

Small Baseline Subset (SBAS) technique is one of the latest algorithms used to monitoring the development of surface displacements over time. The production of time – series of the line-of-sight (LOS) displacement was proposed by Berardion [8]. The SBAS – InSAR use interferograms with a small baseline to minimize topographic errors and the effects of spatial decorrelation – it is the main advantage of SBAS interferometry. Due to the use of distributed scatterers points, which are more commonly identified in the natural environment, this method is used for monitoring displacements in non – urbanized areas. Intermediate and final results (interferograms and time – series displacements maps) are determined by batch processing, in three stages. The advantage of this process is the ability to generate several hundred interferograms based on the large amount of SAR data. All processed and calculations were performed is GMT5SAR.

5.1 Batch pre – processing

The first step before processing includes organize the data. The pre – processing involves both processing without a master image (using default parameters: earth radius, Doppler centroid and near range) and processing with a master image. Command execution sequence is important – the first pre – processing generates the baseline time plot necessary to select the master image to the next step. The best master image covers the area of interest and minimizes perpendicular baselines between interferometric pairs.

5.2 Batch processing

All activities at this stage lead to generation of interferograms. The first step is to select pairs of images from which interferograms are created, specifying the maximum value of the temporal and spatial baseline (Fig. 4). Selection of parameters determining the time intervals and perpendicular baseline should be selected in such a way that the resulting network is consistent and that each image is connected to at least two neighbours. I selected the InSAR combinations that exhibited a mutual perpendicular baseline less than 70 m and corresponding time intervals were set to less than 60 days. With the assumed parameters, a list of 179 interferometric pairs was obtained.

5.3 SBAS – derived time series displacements

This part of the work generates time – series displacement maps (corresponding to the date radar images) and velocity maps. The data parameters were also specified: smooth *sf*, wavelength *wl*, incidence *theta*, range *rng* and number of unwrapped interferograms.

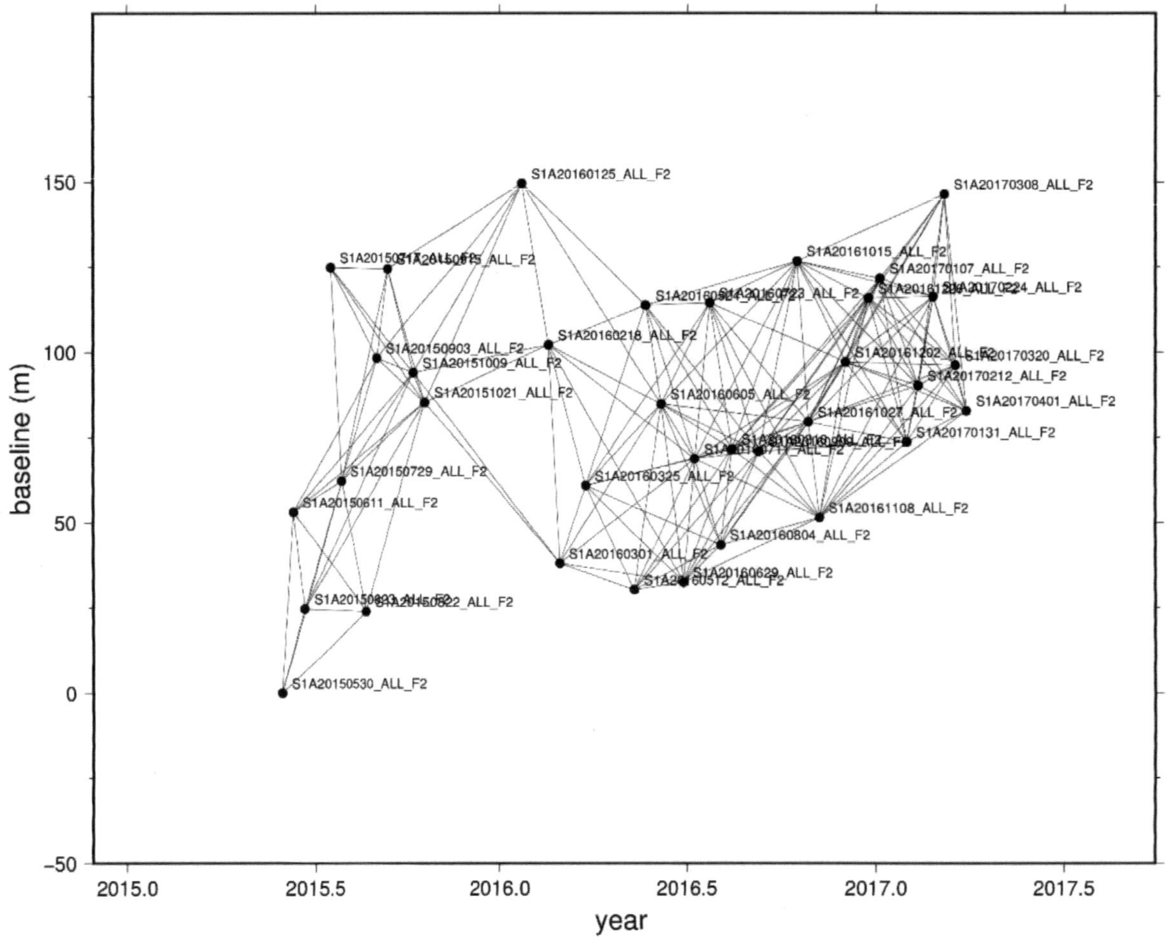

Fig. 4. Baselines and InSAR combinations. Black solid lines connection circle represents perpendicular baselines of the corresponding interferograms.

 https://doi.org/10.1051/e3sconf/20185500006

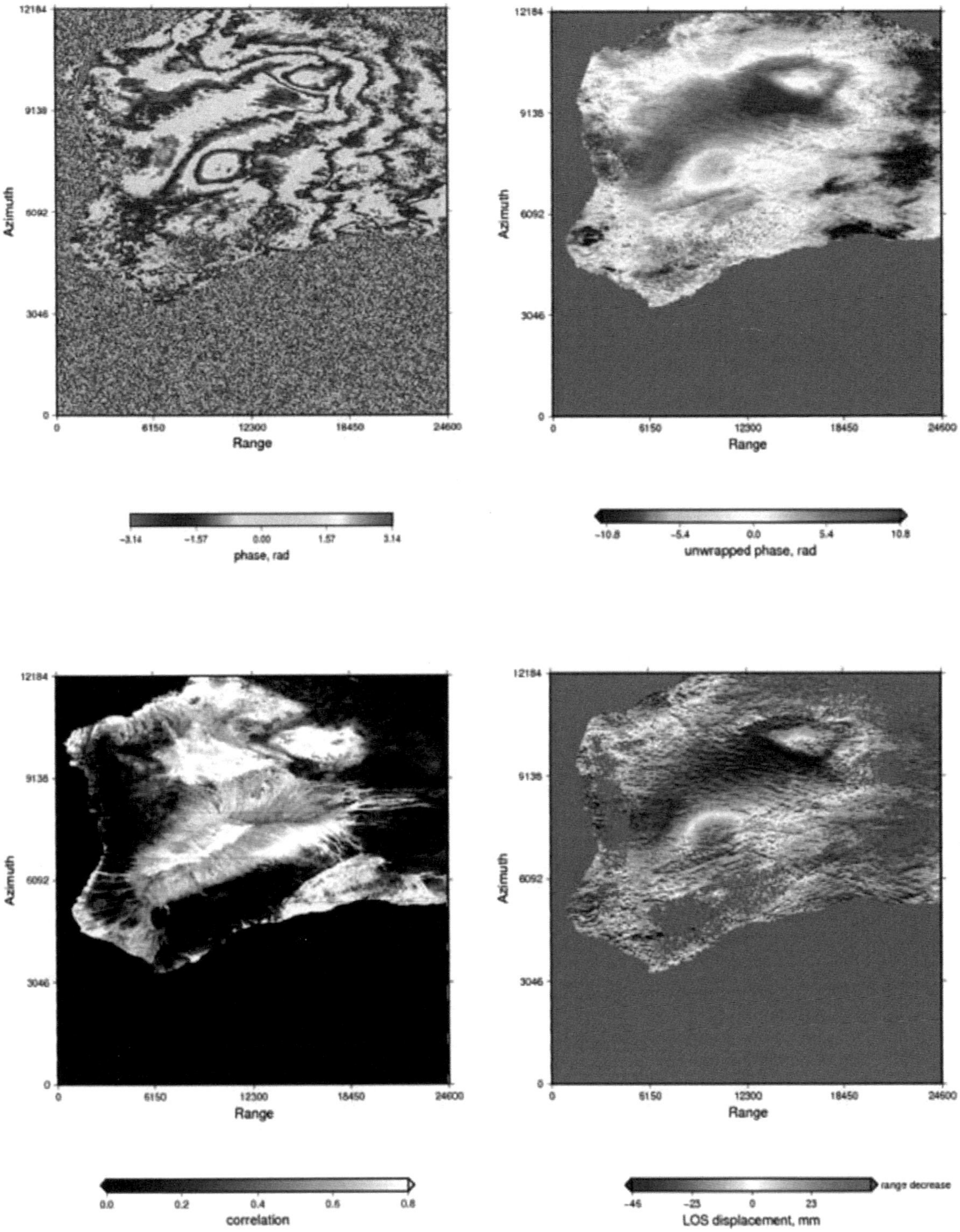

Fig. 5. Examples of calculation results for Hawaii: phase image (upper left), unwrapped phase (upper right), correlation image (lower left) and LOS displacement (lower left).

6 Results

The result of the process of generating interferograms for the first pair of 30.05.2015 and 11.06.2015. The use of the mask caused the elimination of water areas during the generation of interferograms. The resulting images are legible and clearly represent the study area. The correlation map includes the value of the complexity coefficient of the SAR signal, also known as the coherence map. The value is dependent on the phase interference. The value of the coefficient is in the range of 0 to 1, where the pixel receiving values close to equal to 0 exhibit the phase error of the two displays, while the value equal to 1 suggests fill compliance. This example result shows both areas with low and high coherence – these are mainly areas covered by vegetation (for example Forest Reserve areas) is low coherence, but the parts of the volcano surface exhibit high coherence. Based on the coherence map, the areas for which interferometric bands can be produced are determined. Within the first 12 days of observation, recorded displacement value is 4.5 cm and is located near the Kilauea volcano. In relation to the Mauna Loa volcano, the highest value of LOS displacements can be observed in the central part of the volcano – around the crater – about 3 cm. On the other hand, the slope of the volcano decreases. In the next step, the displacement maps were generated in time – series designated for the images input files. The results are in geographic coordinates system. Additionally, three points are identified for the analysis of volcano activity (Tab. 2).

Table 2. The coordinates of designated measurement points.

Point	Longitude	Latitude
P1	-155.7805	19.6161
P2	-155.6494	19.4572
P3	-155.7072	19.2006

Number one was located on the northern slope of the Mauna Loa volcano, the second was located at the western part of the cone, while the third point was on the southern flank of the volcanic edifice, due to the frequency of lava over recent years. As previously mentioned, the Island of Hawaii and volcanoes are observed by the USGS Hawaiian Volcano Observatory. According to published data of 17 September 2017, since mid-2013, seismic activity in the Mauna Loa volcano has increased compared to previous years. Prior to 2013, on average, 15 to 20 quakes were recorded in a week, with low magnitude. However, in 2013 there was a breakthrough. Initially minor changes were observed in the frequency of the occurrence of earthquakes. By the end of 2014, the number of earthquakes rose twice during the week and thrice in 2015 – signalling changes in the earth. Based on a study of the relative distances between the GPS stations on opposite side of Mauna Loa's summit caldera, a value was changed (over 5 years period). It is often an indication of inflation of the shallow summit magma reservoir.

Figure 6 shows the time series Mauna Loa displacements fields from SBAS InSAR. To illustrate the exact displacement values, a graph of the time dependence of displacements values for the three points was created.

The displacement order increased near caldera, and the largest inflating displacement was 34 cm in 432 days (but the largest inflating displacement in the map range was greater than 75 cm in 672 days in around the Kilauea caldera). The changes in the location of the GPS station, according to observers, is due to the magma filling of the complex tank system beneath the peak and the upper southwestern slot. Slow magma movements and filling of the tanks, under the influence of temperature and pressure, cause displacements of the layers and thus displacement of the ground. From time – series displacements maps, it has been observed that from 468 days (09 September 2016) there was a decrease in the area – the lighter colours of the crater in 17g and 17 h Figure 6. The greatest decrease was recorded in the interval of time between 576th day of measurement and 624th – it was 12.5 cm change. From February 2017 until the end of the measurement period, the displacement of the P2 point was observed again.

Displacement at point 1 oscillates within the range of -8 cm to 8 cm, thus not showing too much activity over two years. Maximum displacements between imaging registration dates are estimated at 2 – 3 cm. In the first two months of research, the graph indicates a decrease in ground. By mid – 2016 measured displacement values increase, thus indicating positive displacement. These changes could have been caused by numerous earthquakes, which despite a small amount of power could have influenced the changes in terrain. Point P3 is located in the lower part of the Southwest Rift Zone. The Fig. 7 largely indicates the reduction of the area since January 2016, except for a few intervals where the value of displacement increases (e.g. Between 24th May 2016 and 5th June 2016, the highest value of displacement amounted to 7 cm). In mid – February, the point was increased relative to the previous months and showed a 10 cm increase by the end of the period. In additional, velocity displacement maps were generated as part of the calculations. It represents mean velocity displacement exactly. Fig. 8 shows mean velocity displacement. The displacement values on the maps will be correct when the number of images in each year was approximately equal. The largest annual changes of the analysed object are observed for the Mauna Loa volcano crater, where ground displacement is increasing from 20 to 25 cm per year. The upper south – western slope area shows a decrease of about 10 cm per year, while the lower part of volcano does not show activity – it is stable. Mauna Loa volcanic surface studies were conducted by Casu [3]. From 2003 to 2008 the average annual velocity displacements near caldera was 2.5 cm. It can be stated that volcanic activity has increased over the years.

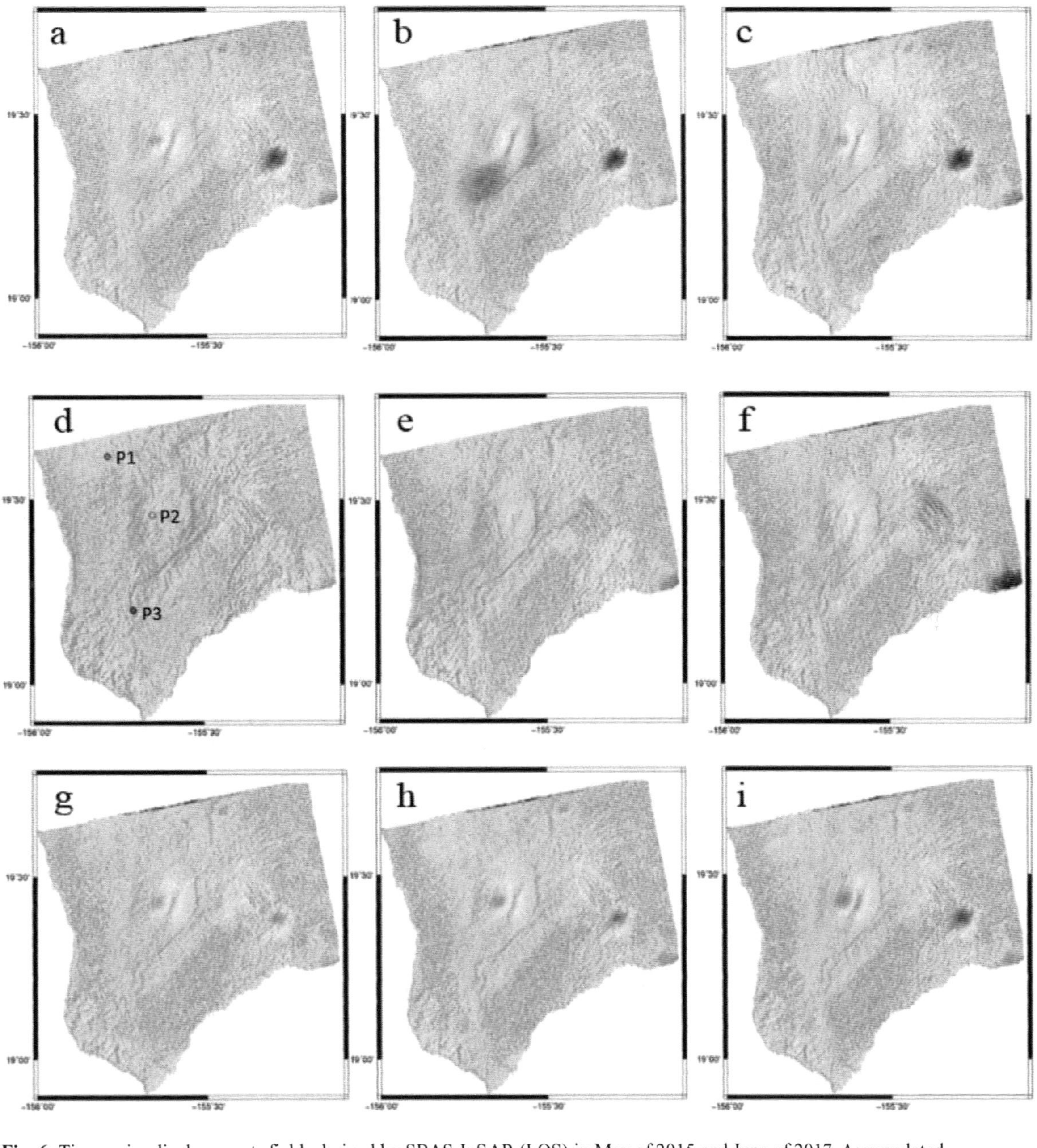

Fig. 6. Time series displacements fields derived by SBAS-InSAR (LOS) in May of 2015 and June of 2017. Accumulated displacements of (a) 24 days (b) 96 days (c) 240 days (d) 348 days (e) 408 days (f) 468 days (g) 552 days (h) 624 days (i) 672 days, are shown.

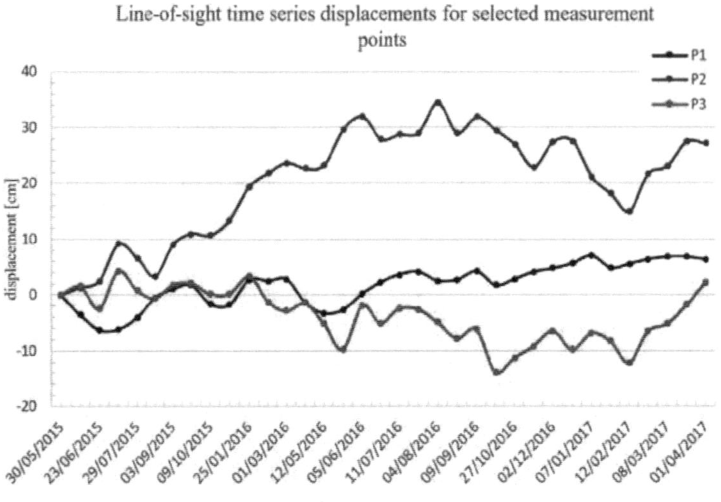

Fig. 7. InSAR LOS time series displacements for selected points.

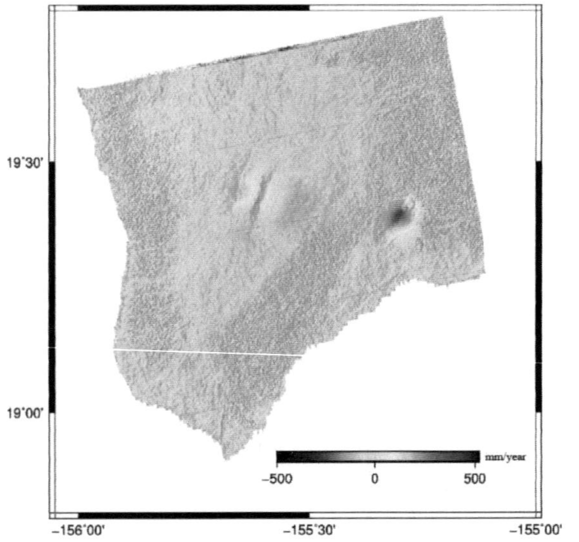

Fig. 8. InSAR LOS time series displacements for entire analysed period.

6 Conclusions

Thirtyfive InSAR images were used to map the displacements of the Mauna Loa volcano in 2015 – 2017. The displacements results showed that the Mauna Loa volcano inflated at the top of the crater. Based on the results, it can be concluded that the algorithm is efficient in determining time displacements on large non – urbanized surfaces. The largest annual changes of the analyzed object are observed for the Mauna Loa volcanic crater. Based on the above results and data from the USGS, it can be said that frequent movements of Mauna Loa volcano surface result from characteristic geological processes there. The lack of eruption at the time of the study does not indicate that volcanic activity is impaired, as earthquakes can occur both on volcanic as well as on tectonic surfaces. Therefore, it is still believed that Mauna Loa is an active volcano. However, in order to obtain information regarding the exact causes of the displacement, measurements should be carried out with a variety of techniques that complement one another. The use of InSAR method in the study of volcanic surface activity is the most correct, because it generates ground displacements maps with a large spatial range of centimetres accuracy. InSAR technology allows you to track ground displacements even in bad weather conditions. Modern measurement techniques allow volcanic activity to be determined, but to be precise about the time and location of an eruption is almost impossible.

References

1. P. Migoń, *Geomorfologia* (Wydawnictwo Naukowe PWN, Warszawa, 2009)
2. D. Sandwell, R. Mellors, X. Tong, M. Wei, P. Wessel, Transactions American Geophysical Union, **92 (28)**, 234, doi:10.1029/2011EO280002 (2011)
3. F. Casu, R. Lanari, E. Sansosti, Geoscience and Remote Sensing Symposium, IGARSS. Volume: **IV**, s. 41-44 DOI: 10.1109/IGARSS.2009.5417600 (2009)
4. F.A. Trusdell, *Mauna Loa – History, Hazards, and Risk of Living With the World's Largest Volcano* (USGS Fact Sheet 2012-3104, 2012)
5. F. Amelung, S. Yun, S., T. Walter, P. Segall, S. Kim, Science **316(5827)**, 1026-1030 (2007)
6. R. Tilling, C. Heliker, D.A. Swanson, D.A., *Eruption of Hawaiian Volcanoes – Past, Present, and Future* (U.S. Geological Survey General Information, 2010)
7. T. Farr, P. Rosen, E. Caro, R. Crippen, R. Duren, S. Hensley, M. Kobrick, M. Paller, E. Rodriguez, L. Roth, D. Seal, S. Sha_er, J. Shimada, J. Umland, M. Werner, M. Oskin, D. Burbank, D. Alsdorf, Reviews of Geophysics, **45** (2007)
8. P. Berardion, IEEE Transactions on geoscience and remote sensing, Volume **40**, Issue 11, 2375-2382 (2002)

Changes in the conduct of administrative proceedings in the scope of cadaster database

Ewa Sudoł[1,*]

[1]Wrocław University of Science and Technology, Faculty of Geoengineering, Mining and Geology, 27 Wyb. Wyspiańskiego St., 50-370 Wrocław, Poland

Abstract. One of the components of the national spatial information infrastructure database (National Geographic Information System) at the district level is the records database of land and buildings. It is a state register that contains databases related to space and procedures and techniques for collecting, updating, processing and sharing information. A legal system regulating the procedures for keeping, updating and making available subject and object information included in the cadastres database is determined by the proper functioning of the land and buildings register (as part of the geodetic resource). The basis of department work in the field of keeping land and buildings records are primarily provisions of the Codex of Administrative Procedure of the Geodetic and Cartographic Law and regulations in the register of land and buildings. On 1 June 2017, a new Codex of Administrative Procedure came into force. In this paper have been presented the most important changes in conducting of administrative procedures in the field of geodesy at the district level after the aforementioned legal amendment. This paper touched the topics related to, among others, issued principle of resolving doubts in favor of the party (Article 7a), silent settlement of the case (Article 122a), administrative simplified proceedings (Chapter 14), appeal proceedings (Chapter 10). There have been presented examples of the most frequently conducted cases in the field of land and buildings registration in district country with particular emphasis on changes introduced by the latest legal amendments in connection with detailed regulations. The author focused on the logical interpretation of legal changes in order to show the problems connected with their application in practice (from the point of view of the proceedings and local government administration).

1 Law basis for updating the cadaster database

Spatial information refers to the location of the object field. The part of this data used by individuals and institutions, for example for investment purposes and ownership management. For the area of the country, a database is established and maintained in the ICT system, including spatial data sets of the spatial information infrastructure, regarding land and building registers. Data sets collected in databases constitute the basis of the national system of information about the area, which is a component of the spatial information infrastructure referred to in Article. 3 point 2 of the Act of March 4, 2010 on the infrastructure of spatial information [1]. The part of the information is collected and stored in the Country Center for Geodetic and Cartographic Documentation. These data allow public entities, entrepreneurs and citizens with easy access to reliable, complete and current information about real estate.

The proper functioning of data and its continuous use forces the need for continuous updating. This process is governed by the provisions of the Geodetic and Cartographic Law Act and the Ordinance of the Minister of Regional Development and Construction regarding land and building registers, stating that it is the task of the mayor to keep records to maintain the inventory in the state of news that is, compliance with the documents and source materials available to the body and periodic verification of registration data. Changes in databases are carried out ex officio or at the request of a party. An example of proceedings initiated at the request of the parties may be references to changes made to the database of land and buildings as part of the modernization (allegations of the entered data) conducted on the basis of article 24a of the Act [1], changes introduced on the basis of technical operators prepared by authorized surveyors (in the absence of the owner's signature), refusal to release data from a descriptive part of the land and building registry database or to enter data into the database or in the field of the classification of land Soil. Proceedings conducted ex officio at district level are most often carried out in the scope of changing this information resulting from legal regulations, materials of the resource, and detection of erroneous information, at the request of the owners or owners of land on the principle of self-possession. The legal basis of the proceedings is primarily Article 22 section 1 of the Act of May 17, 1989. Geodetic and

* Corresponding author: ewa.sudol@pwr.edu.pl

cartographic law (Dz.U.2016.1629 i.e., 2016.10.06 with amendments) [1], 47 paragraph 3 of the Ordinance of the Minister of Regional Development and Construction from March 29, 2001. on the registration of lands and buildings (Dz.U.2016.1034 from 2016.07.15) [2] and Article 104 of the Act of 14 June 1960. The Code of Administrative Procedure (consolidated text, Dz.U.2017.1257 from 2017.06.27) [3]. In accordance with the judgments of the Supreme Administrative Court of November 19, 1999, reference number II SA1384 / 99, dated 20 August 1998, Ref. II SA 766/98, the provisions of the Geodetic and Cartographic Law Act do not provide legal grounds for shaping ownership relations and related rights and obligations. The body running the custodial survey is not entitled to settle matters relating to property ownership [4].

The information contained in the land and building register is updated by means of material and technical activity on the basis of legal regulations, entries in land and mortgage registers, final court decisions, and in cases related to the European Certificate of Succession - court rulings, final administrative decisions, notarial deeds, certificates of inheritance and European declarations of inheritance, building declarations, entries in other public registers, the application of the interested registration entity and the geodetic documentation indicated in the application, adopted to the state geodetic and cartographic resource, if the change requested includes information collected in the land and building register for real estate in the exclusive jurisdiction of the applicant or applicants, by way of an administrative decision - in other cases (Article 24) [1]. The legal foundation for administrative proceedings regarding the updating of cadaster databases is the amended Code of Administrative Procedure. The provisions of the above-mentioned act normalize the work schedule carried out in district offices (Fig. 1).

The result of the high detail of decision bases is their high volatility, which forces the construction of websites to ensure a systematic and continuous data update. The Spatial Information Infrastructure Act, which is - inter alia - the implementation of the EU directive "INSPIRE" into Polish law, obligates to provide IT access to databases (including geodetic databases) [5]. At the same time, it enforces the computerization of the resource. This way of providing spatial information means that consumers of this information become verifying its correctness, compliance with the actual and legal status as well as timeliness. This, in turn, enforces high quality and credibility of the data.

2 The time of handling cases. Urgency as a means of challenging inaction and excessive length

The main objectives of the amendment of the CAP were to streamline and speed up administrative procedures, create a framework for more partner relations between citizens and the administration and ensure the adequacy of administrative penalties for violation of the law. The basic expectation related to the amendment of the law resulted from the need to shorten the time of handling cases.

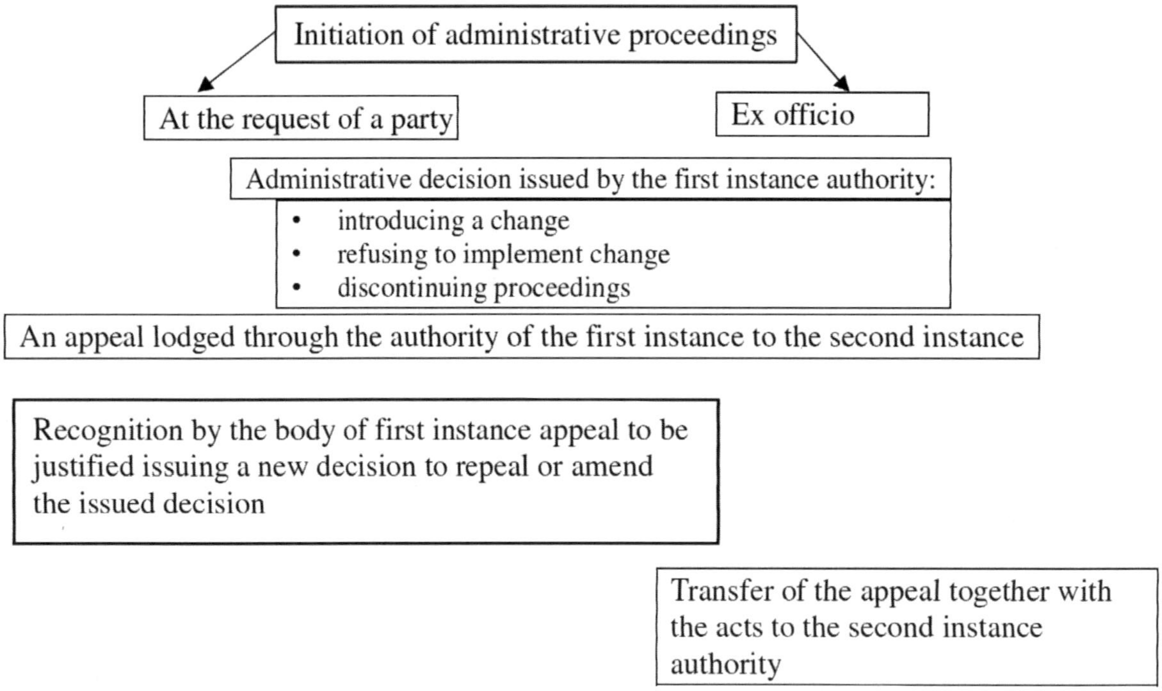

Fig. 1. Diagram of administrative proceedings following the amendment of the CAP.

The amendment to the Administrative Procedure Code introduced the following modifications:

- inactivity of organs or excessive length of proceedings can be challenged in a transparent procedure by means of a reminder (Fig. 2),
- a request for a retrial is optional - without using it you can make a complaint to the administrative court,
- the party may request a second instance authority to issue a decision deciding the case instead of the so-called a cassation decision, that is, forwarding the case to be reconsidered,
- a cassation decision may be appealed to the administrative court in a special accelerated procedure,
- the decision can become final before thanks to the waiver of appeal,
- the framework was introduced procedure of tacit deal with the party's request and simplified procedure, which would reduce formalism in simple cases.

3 New administration utility

Pursuant to the amended Code of Administrative Procedure, from 1 June 2017, no appeal to the Provincial Administrative Court can be made against a cassation decision. The amendment of the CAP granted a new appeal to the party - an objection to the decision (Fig. 3).

If in the case there are grounds for issuing a cassation decision, the second instance authority may refer the case back to the first instance. The guidelines of the second instance authority are not formally binding for the first instance authority. An interpretation of the relevant provisions after a detailed determination of the facts of the case and their application remains the responsibility of the first instance authority. Due to the change, if the authority of the second instance notices a mistake in interpreting the law, the first-instance authority will be able to avoid it by reconsidering the case. Another change that has a significant impact on the date of consideration of the case is undoubtedly the possibility of waiving the right to appeal against the decision in accordance with article 127a CAP. It means that after receiving the decision and using this utility, for example, the investment process can be started more quickly, after the division decision has been received, geodetic works can be carried out without having to wait for the final decision.

A silent solution to the matter is a new utility to accelerate and facilitate the course of proceedings. It lies in the fact that the "silence" of the administrative authority, i.e. no decision after the deadline for settlement of the case shall be deemed to settle this matter in a way that takes into account the whole page request. Silent settlement of the case is not applicable in all proceedings. Only the provision concerning a specific administrative procedure will be the basis for a silent settlement of the matter. The CAP [3] amendment introduced a model procedure in which this will take place at the time when the detailed provisions appear. In the field of geodesy and cartography are no specific guidelines for division offices, which would allow for the use of this utility. A similar legal status applies to the mediation utility. Administrative mediation can be carried out in all matters, the nature of which allows it, i.e. in cases in which there are parties with disputed interests, i.e. Neighborhood disputes or long-lasting disputes between the parties, but also in cases in which we have to deal with a complicated state of affairs. Unfortunately, the use of this tool also requires the adoption of specific provisions.

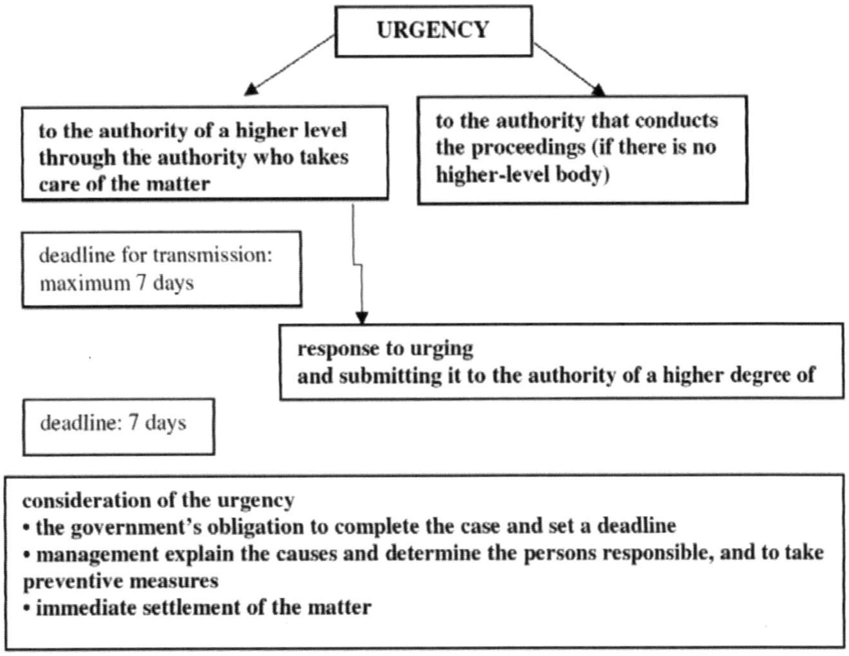

Fig. 2. Diagram of the urgency procedure.

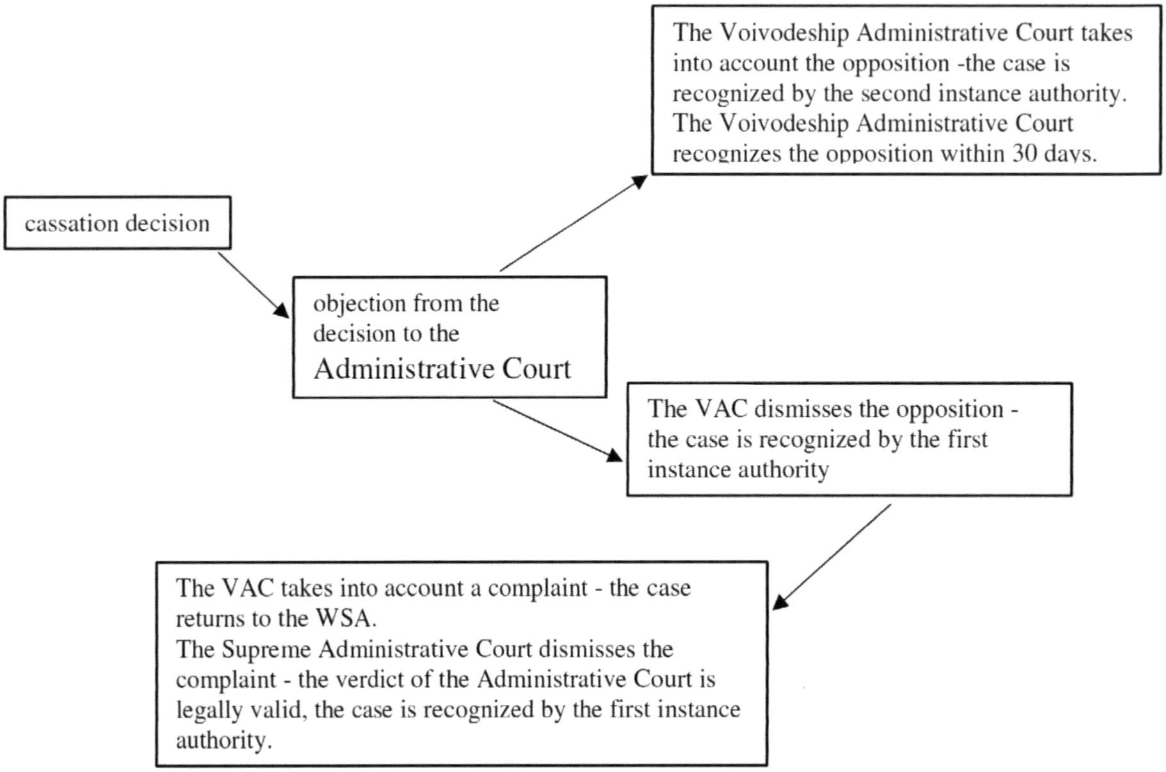

Fig. 3. Opposition to the decision - the course of proceedings [6].

4 Conclusions

A modern geodetic resource is based on data bases established and maintained in the ICT system for the entire country. Data update procedures should be as streamlined as possible and as short as possible. Only valid and reliable spatial information allows the use of modern forms of resource sharing, but also new forms of content enrichment services such as high-speed access to data in the resource, the communication process automation contractor - administration. The legal tools described in the amendment to the Code of Administrative Procedure give the opportunity to shorten the time of proceedings related to the updating of geodetic databases. Unfortunately, the current legal status is not complete and specific provisions incompatible with new administrative utilities. There are not precisely formulated range of cases that could be resolved using mediation or in silent mode on your behalf or having simplified administrative proceedings. Experts point out that the above changes were passed very quickly (the government passed them on 7 April), which up to now causes problems with the application of the new regulations.

References

1. Ustawa z dnia 17 maja 1989 r. Prawo geodezyjne i kartograficzne, Dz.U.2016.1629 t.j. z dnia 2016.10.06
2. Rozporządzenie Ministra Rozwoju Regionalnego i Budownictwa z dnia 29 marca 2001 r. w sprawie ewidencji gruntów i budynków, Dz.U.2016.1034 t.j. z dnia 2016.07.15
3. Ustawa z dnia 14 czerwca 1960 r. Kodeks postępowania administracyjnego, Dz.U.2017.1257 t.j. z dnia 2017.06.27
4. Wyroki NSA z dnia 19 listopada 1999r., sygn. II SA1384/99, z dnia 20 sierpnia 1998r., sygn. II SA 766/98
5. Nowoczesny Zasób Geodezyjny i Kartograficzny, Wojewódzka Inspekcja Geodezyjna, Kraków (2015)
6. Nowelizacja Kodeksu Postępowania Administracyjnego, Czego dotyczą zmiany i jak je stosować? Ministerstwo Rozwoju, maj 2017 r.

Application of new measurement technology for deformation study of structures in mining areas

Piotr Gruchlik[1,*], and *Andrzej* Kowalski[1]

[1]Central Mining Institute, Department of Surface and Structures Protection, 1 Gwarków Sq., 40-166 Katowice, Poland

Abstract. Despite of classic geodetic measurements, which are used in mine surveying at the beginning, more new technology are applied to measure deformation of structures and surface in mining areas. In the article new technologies are described for utilitarian aims based on strain sensors, laser, scanners and drones. The application of new technologies are especially useful when direct access to measured object is impossible using classic methods and in situation where continuous or quasi-continuous measurement is necessary. In Central Mining Institute the research has been conducted to understand the phenomenon of surface deformation induced by underground mining comparing predicted and measured values before and after extraction. In the last 20 years the mining technology has changed and dynamics of structure and surface deformation induced by mining. The second reason of technology development is necessity to instantly inform users of land and structures about changes using for instance the Internet.

1 Introduction

New survey technology (sensors, inclinometers, extensometers, laser, terrestrial and aerial laser scanning - LIDAR – Light Detection and Ranging) extents monitoring of surface deformation. However, these technologies can't measure everything because they have some limitations. The ideal method doesn't exist.

Geodetic surveys still are basic and reference methods, which give information about coordinates of points on the bare-earth and structures. They may also give information about changes in time. Development of mining technology induces increasing deformation of surface land and structures and includes sizeable area. Second aspect that enforces application of new survey technology is reaction of citizens on negative results of underground mining. The users of structures and land expect instantly information about deformation and warnings of usage risk.

In Central Mining Institute the research has been conducted to understand the phenomenon of surface deformation induced by underground mining comparing predicted and measured values before and after extraction. It's significant because without survey the knowledge is blind [1].

In recent years in Central Mining Institute new survey technology has been used to monitor of sacral and industrial buildings as supplementation of geodetic measurements.

Interpretation of both ways to store data enables to better understand the phenomenon of deformation. Correctly designed monitoring of deformation process should include not only way of getting data but also allow for correct interpretation. The results should give the answers for questions such as: will goals be achieved, how to change the mining and structures prevention and is changing survey method is necessary?

2 New survey technology

2.1 Extensometers and inclinometers

Extensometers and inclinometers are applied to measure deformation of structures in mining areas. They are usually connected with others devices, which play function as receiver or transmitter or both. Setting of suitable system with algorithms may enable to provide automatic measurements for long time. Wireless monitoring system of structure construction includes:

- string extensometers to measure linear strains or width of cracks,
- string inclinometers to measure three dimensional inclination of building.

First instruments were constructed in The Strata Mechanics Research Institute of the Polish Academy of Sciences [2]. The system has been applied for remote monitoring of building construction of Lutheran Church in Bytom-Miechowice (compare section 2.2) induced by longwall 18a in lower layer of 510 seam extracted during 2011-2012, Fig. 1 [3]. In this church cracks on walls and roofs have emerged. The construction of the building has been induced by uneven subsidence.

Surveys proved that system was working well to monitor deformation of construction of the church building during deformation of ground induced by mining in terms of time. Three dimensional surveys of

* Corresponding author: p.gruchlik@gig.eu

the church tower enabled to indicate main directions of inclination.

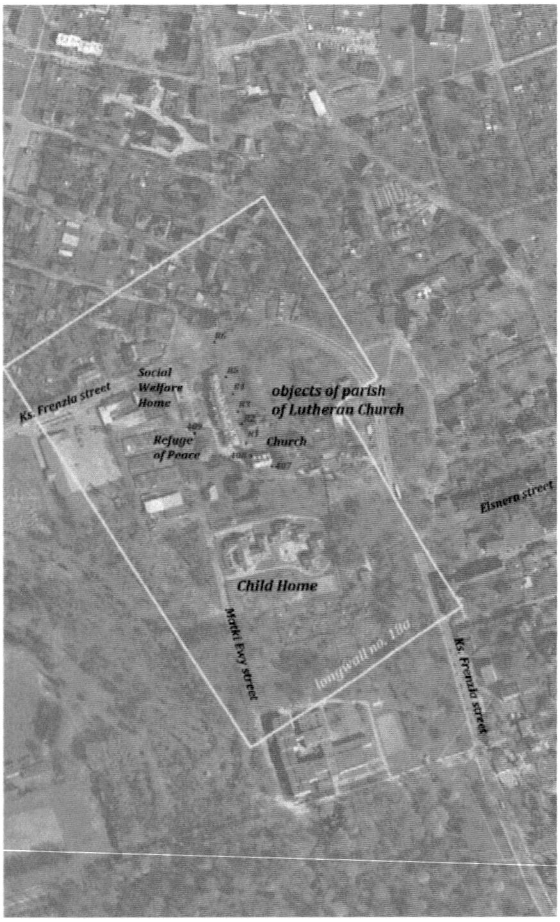

Fig. 1. The fragment of the measured power line and the set measuring instrument [3].

Sensitivity of the system enables to monitor small changes of strains. Disadvantage of the system is the scattered observation status. Considering small number of sensors there is risk to miss some significant information about a reaction of building construction. Despite of numerous advantages, the remote system cannot be the only one tool to monitor a condition of the object.

Another but similar system has been applied in two viaducts (road and tram viaduct) in Ruda Śląska (Fig. 4). The monitor system includes twelve sensors as follows:

- eight crackmeters (Geokon 4420) to measure displacement increments of deck in term of abutments along y axis,
- four vibrating wire tiltmeters (Geokon 6350) to measure inclination of abutments along two perpendicular planes and 16-channel recorder to store data.

First results of the measurement have been read 30 days time. Later monitoring system was extent by next four sensors. In the meantime, the system of storing and processing data has been replaced with new one to measure tilts and displacements in real time.

Received displacement increments of road viaduct abutments in real time were similar to predicted value. Whereas, displacement increments of tram viaduct were completely different than the predicted values. It may be interpreted as viaduct work.

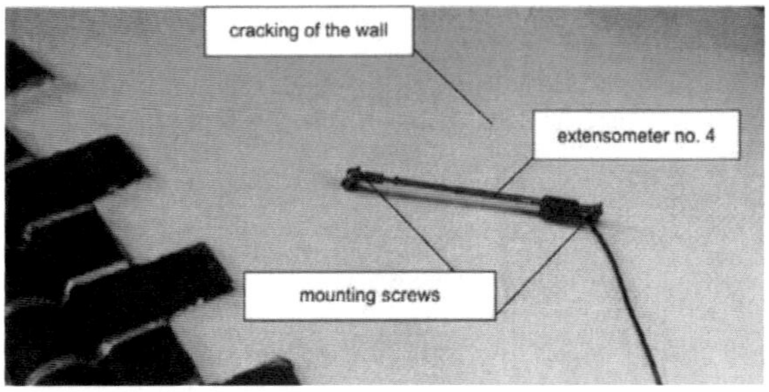

Fig. 2. Installation of extensometer number 4 over the crack of the wall nearby main entrance of the church [4].

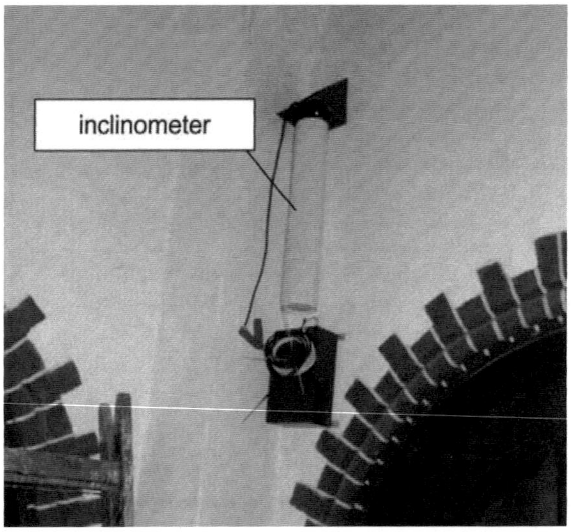

Fig. 3. Installation of inclinometer in quoin of the church tower [4].

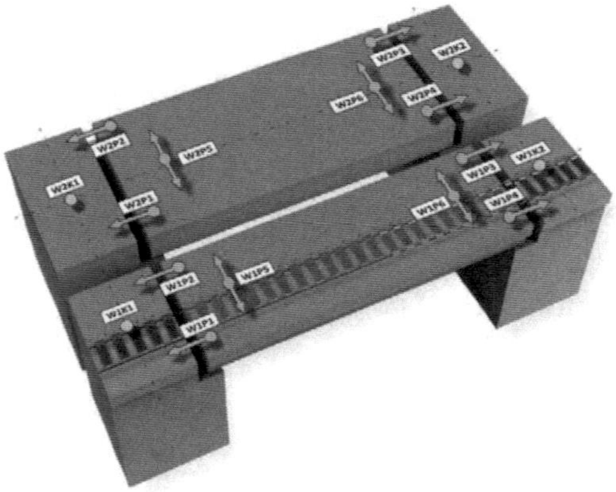

Fig. 4. Arrangement of sensors on tram and road viaduct designed to measure displacement increments (WP) and arrangement of inclinometers (WK).

2.2 Extensometers and inclinometers

The discussed laser sensors for inclination and acceleration measurements are designed by Central Mining Institute. Sensors provide accurate and continuous work to measure inclination of structure. The output data can be shown on the graphical chart or recorded in memory of decoder including values of inclination (millimeter per meter unit), directions of inclination (e.g. related to cardinal directions) and time of registered data. Principle of survey is based on recording track of laser beam from vertical plane after passing through liquid wedge.

Specification of sensor is as follow:

- defined measurement range +5 mm/m mm/m, (range of sensor 2-50 mm/m),
- resolution: +0.02 mm/m,
- frequency of vibration <10 Hz,
- power supply voltage of device: 220 V A/D, 50/60 Hz,
- power consumption <2 W.

Recording system consists of multichannel A/D converter with resolution 12-bit, PC computer and hard disc drive. Controlling of recorder, viewing and transmission data are able trough website using GPRS technology.

Inclination sensor and digital recorder were installed in steal covers on south-east side of "Terma-Dom" chimney base about 2 m above the floor in Zabrze (Fig. 5). The chimney was described in details in section 2.3. The monitoring system has been started on the 22.03.2016. The inclination directions of sensor were calibrated according to cardinal directions. Before field use it had been tested in the laboratory on swing table with micrometer screw.

In the first period of measurements inclination increments of chimney were affected by atmospheric factors (temperature, wind, sun exposure). Temporary chimney inclination of about +0.05 mm/m has occurred because of wind pressure (average wind speed equalled 7 m/s).

Next, daily increments of chimney inclination are shown in Fig. 6. Fig. 7 shows values of resultant vector in plane X – Y (S-N – W-E). Based on the mentioned charts, temperature (and sun exposure) above 10°C triggers chimney inclination in south-west direction (-0.3 mm/m south for temperature 25°C). Temperature below 10°C induces inclination in opposite direction (0.4 mm/m north for temperature 1°C). Between 28 of June to 8 of July direction of chimney inclination was south-east, whereas from 14 to 31 of July rapidly had been change to south-west. The highest values of vector components (1.3 mm/m west and 1.89 mm/m south) has been recorded between 4 and 5 of August.

Fig. 5. The place of sensor installation.

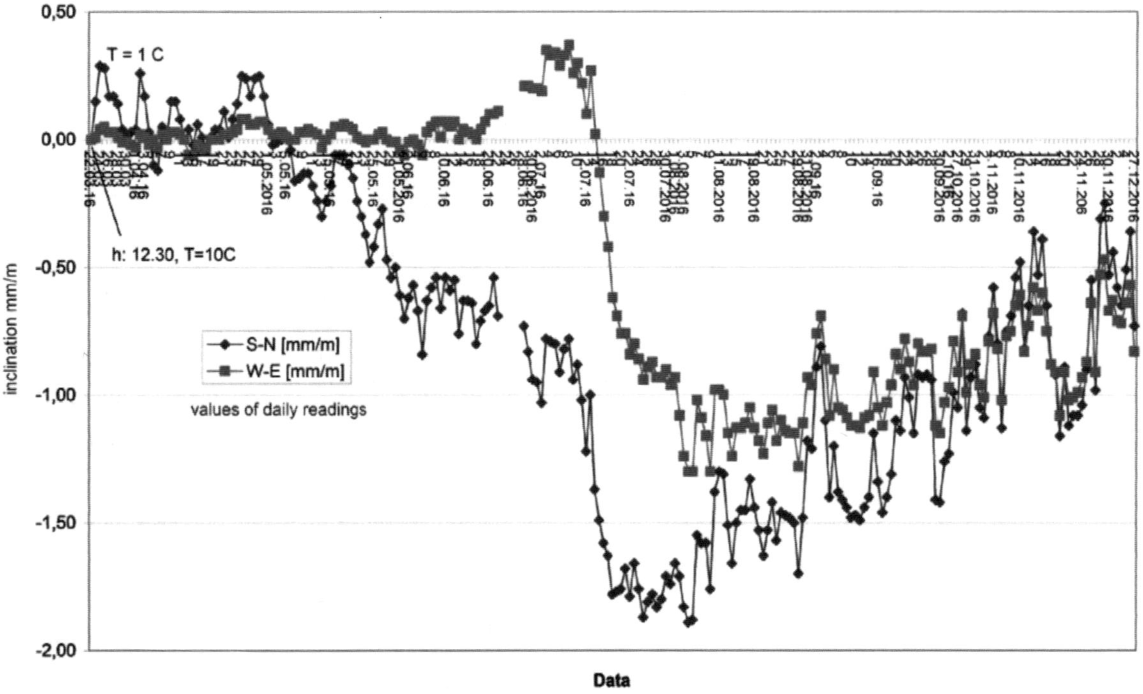

Fig. 6. Inclination distribution of chimney base along W-E and N-S axes.

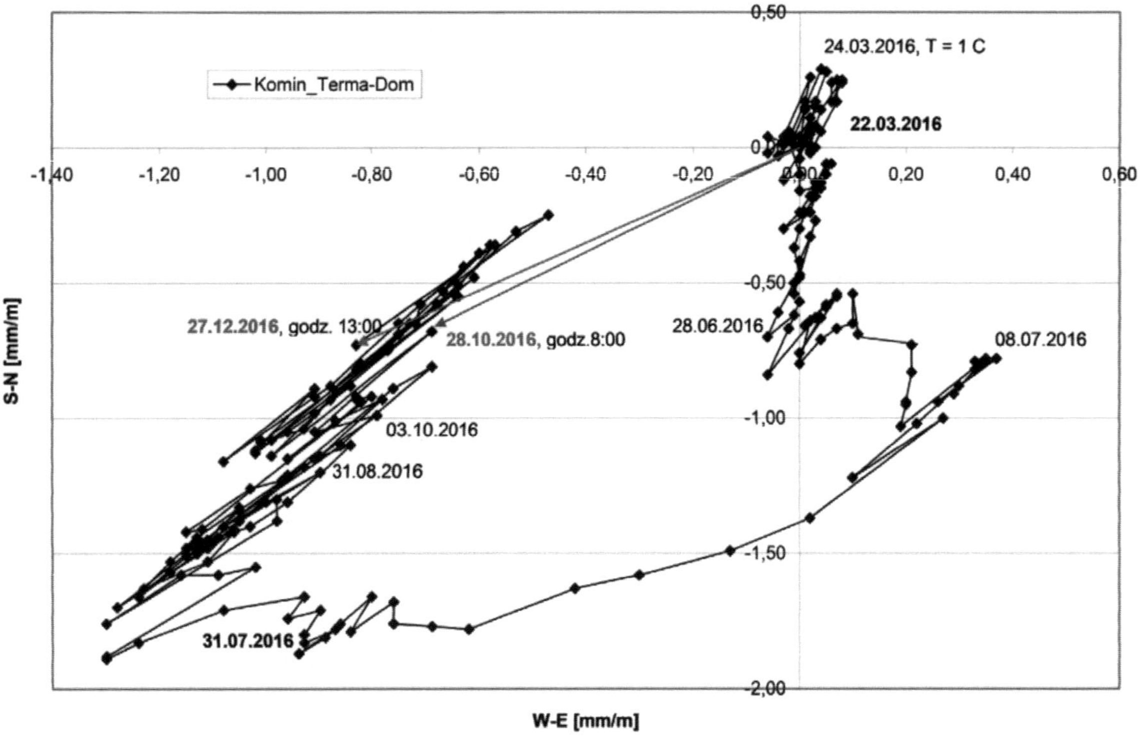

Fig. 7. Inclination vectors of chimney base since the day of sensor installation.

The calculated length of resultant vector is 2.3 mm/m in south-west direction according to the first day of storing data. In September 2016 there were periodic sinusoidal changes of direction of chimney inclination (+0.3 mm/m). The resultant vector has been decreased by 0.2 mm/m in north-east direction. This tendency has been remained until mid-November and the value of resultant vector dropped to 0.72 mm/m. In December increments of chimney inclination were around mean value of increment of resultant chimney inclination (1.0 mm/m south-west) according to the date of first survey.

2.3 3D laser scanning

Department of Surface and Structures Protection in Central Mining Institute has 3D laser scanner Trimble TX5, which enables to store data with speed 976 000 points per second at a range of up to 120 m [5, 6]. For high objects such as chimneys it is possible to use 3D laser scanner Riegl VZ-4000 too (Fig. 8).

Usage of laser scanning technology enables to asses mining activity on structures and asses their resistance to mining deformation of the ground.

The object of scanning was building of Lutheran Church situated on Matki Ewy Street in Miechowice (section 2.1, Fig. 9). Indoor surveys were carried out using 3D Trimble TX5 scanner. The cracks of the church roof are shown in Fig. 9.

With 3D Trimble TX5 laser scanner it's easy to collect data of actual condition of object. The data can be used to design, redesign, and for technical review of structure. It is more convenient to compare the design and as-built state of the building too. Usage of the scanner is wide for example to get cubature of underground workings, dumps and other object with irregular shape.

The second field of scanning was in the area of coke plant placed in Zabrze-Biskupice where the "Siltech" mine extracts coal seams using experimental method leaving the coal pillars between backfilled headings. Fig. 10 shows extracted panels with headings system, observed subsidence during the 2015-2017 period and coke oven battery, 113 m high "Terma-Dom" chimney, which was captured using the above-mentioned 3D laser scanner on 09.06.2017.

The results of measurements were points cloud and model of chimney (Fig. 11). cross-sections were made to analyze the body of chimney (Fig. 12). Calculated inclination of top of the chimney based on scanning is equal to 575 mm with azimuth 33° in mathematical system and 57° in geodetic system (north-east direction).

Inclination surveys of top o the chimney are carried out by geodetic firm as result of a contract with the owner of the chimney. The first and subsequent measurements have captured the state of this object before and during mining under investigated area. Resultant inclination of top of the chimney is 63 cm in geodetic system (from north to clockwise). Mean error of inclination equals ±3,0 cm.

Fig. 8. 3D laser scanner Riegl VZ-400.

Fig. 9. 3D model of church building based on indoor surveys, north side view [6].

Fig. 10. Extraction of panels and headings in seams 507, 509 and 510 with subsidence contours.

Fig. 11. Registered cloud of points and model of Terma-Dom chimney.

Tables 1 and 2 include inclinations of top of the chimney using geodetic and scanning methods. Boundary condition of strength based on material strength of top of the chimney is related to horizontal displacement of up to 70 cm.

Historical and up-to-date results of surveys are in Tables 1 and 2. In 2016 the chimney has been affected by mining that resulted in inclination increments up to 6 cm, according to geodetic method and up to 5 cm according to the scanning one.

Fig. 12. Horizontal cross sections of 3D chimney model.

Table 1. Inclinations of top of chimney using geodetic methods.

Number of survey	Date of survey	Inclination	Azimuth of inclination
1	IV.2011	57	49°51'
2	X.2011	56	52°24'
3	XII.2011	55	55°49'
4	IV.2012	57	55°21'
5	X.2012	59	55°34'
6	IV.2013	55	52°16'
7	X.2013	60	56°56'
8	IV.2014	55	49°32'
9	X.2014	56	51°43'
10	III.2015	56	50°47'
11	VII.2015	57	50°47'
12	IX.2015	58	46°01'
13	IV.2016	58	51°31'
14	29.IX.2016	64	55°27'
15	31.X.2016	63	57°00'

Table 2. Inclinations of top of chimney using laser scanning.

Number of survey	Date of survey	Inclination	Azimuth of inclination	Type of scanner	Weather condition in a day of survey
1	29.07.2015	56	55°	Trimble TX8	27°C, 7,2 km/h
2	04.11.2015	56	51°	Trimble TX8	lack of data
3	18.03.2016	55	54°	Trimble TX8	10°C, W 34 km/h
4	21.06.2016	54	64°	Riegl VZ-4000	21°C, SW 14 km/h
5	26.09.2016	56	58°	Riegl VZ-4000	18°C, E,EW 10 km/h
6	18.10.2016	59	58°	Riegl VZ-4000	11°C, SSW 18 km/h
7	22.12.2016	58	60°	Riegl VZ-4000	-1°C, SSW 20 km/h
8	09.06.2017	57.5	57°	Riegl VZ-4000	23°C, SE 9 km/h

Another example of laser scanning applications is related to the building of Saint Cross church in Bytom-Miechowice, Fig. 13. The aim of measurement was to determine curvatures of masonry construction as an indicator of effort construction.

Fig. 13. Building of St. Cross church in Bytom-Miechowice.

Vertical curvatures of the church walls (Fig. 14) were calculated on the basis of points coordinates along particular lines in plane of the walls (Fig. 15).

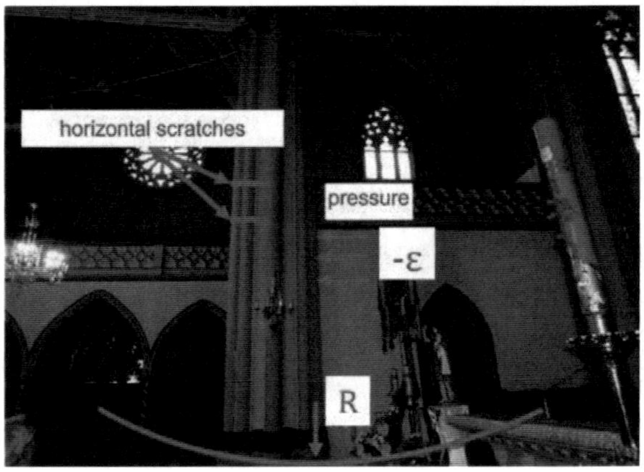

Fig. 14. Scheme of interaction of deformation indicators on construction of church building.

Fig. 15. Horizontal lines in 3D model of the church building to determine curvatures.

2.4 Unmanned Aerial Vehicles

In last year (2016-2017) drones are more popular to review the condition of inaccessible parts of structure construction. Among others they are used to review the condition of brickwork of high chimney Terma-Dom (section 2.3) and damages of the Saint Wawrzyniec church building in Ruda Śląska-Wirek.

Drone is a popular term of unmanned aerial vehicle. It's controlled by qualified operator from ground using remote system and can be programmed to fly along designed trail.

In Central Mining Institute two types of drone are in service, the larger one Quadrocopter Phantom (Fig. 16) and the smaller one scout DJI Mavic Pro (Fig. 17).

Drones are designed to record photo and video material. They can be used in cartography to make orthophotomosaics and to review the construction of structures, power networks, etc.

The review results of 113 m high chimney (application of Quadrocopter Phantom drone) and church building (application of DJI Mavic Pro drone) are shown sequentially in photographs 1-7.

Fig. 16. Drone Quadrocopter Phantom.

Fig. 17. Drone scout DJI Mavic Pro.

Photo 1. View of Jadwiga coke plant and structures of heat station with Terma-Dom chimney.

Photo 2. Top of Terma-Dom chimney.

Photo 3. View of St. Wawrzyniec church in Ruda Śląska - Wirek.

Photo 4. View of the church tower, close-up on the upper part of the tower for inspection of damages in the wall.

Photo 5. View of the western wall of the church nearby tower (cemetery side), visible fissures and rosette deformation.

Photo 6. View of the western wall of the church nearby transept (vertical fissures).

Photo 7. The Eastern wall of the church nearby tower (fissures and rosette deformation).

3 Conclusions

Experience from measurement sessions with application of laser and string sensors to observation of structures deformation indicates their expediency as supplement of classic geodetic methods. Their advantages are continuous and real-time observation, remote transmission of data and signalling of exceeding threshold values.

Terrestrial laser scanning is a source of accurate data to monitor condition of structures and review of mining damages. 3D mapping enables further operation including analyze of measurement results and modeling. Laser scanning brings new quality for monitoring and inventory of mining damages.

Application of drones enables to observation of parts of buildings, which are inaccessible for classic geodetic methods.

References

1. A. Kowalski, *Deformacje powierzchni w Górnośląskim Zagłębiu Węglowym* (Wydawnictwo Głównego Instytutu Górnictwa, 2015)
2. A. Kanciruk, Arch. Min. Sci., **15** (2012)
3. L. Klabis and A. Kowalski, Przegląd Górniczy, **8**, 62-70 (2013)
4. L. Florkowska and A. Kanciruk, [In:] Ochrona obiektów na terenach górniczych (Wydawnictwo Głównego Instytutu Górnictwa, Katowice, 2012)
5. P. Gruchlik, Przegląd Górniczy, **5** (2015)
6. T. Niemiec and P. Gruchlik, Geoinformatica Polonica, **15**, 103-112 (2015)

Approximation of the process of changes in deformation of land surface using artificial neural networks

Maria Mrówczyńska[1,*], and *Jacek* Sztubecki[2]

[1]University of Zielona Góra, Institute of Building Engineering, Faculty of Civil Engineering, Architecture and Environmental Engineering, 1 Szafrana St., 65-516 Zielona Góra, Poland
[2]University of Science and Technology, Faculty of Civil and Environmental Engineering and Architecture, 7 prof. S. Kaliski St., 85-796 Bydgoszcz, Poland

Abstract. Artificial neural networks are an interesting method for modelling phenomena, including spatial phenomena, which are difficult to describe with known mathematical models. The properties of neural networks enable their practical application for solving such problems as: approximation, interpolation, identification and classification of patterns, compression, prediction, etc. The article presents the use of multilayer feedforward artificial neural networks for describing the process of changes in land surface deformation in the area of the Legnica-Głogów Copper Mining Centre, located in the southern part of the Fore Sudetic Monocline. Results provided by geodesic monitoring, which consists of land surveying and interpreting data obtained in this way, are undoubtedly significant in terms of identifying the impact of mining on the land surface the results of measurements carried out by precise levelling in the years 1967-2014 were used to determine changes in land deformation in the Legnica-Głogów Copper Mining Centre. The concept of a flexible reference system was used to assess the stability of points in the measurement and control network stabilized in order to determine vertical displacements. However, the reference system itself was identified on the basis of the critical value of the increment of the square of the norm of corrections to the observations.

1 Introduction

A commonly used method of mathematical description of processes and phenomena is linear modelling, which brings favourable results because it provides a good description of the optimization strategy used in models of this type. However, when there is no basis for the use of linear approximation, and non-linear models bring unfavourable results, artificial neural networks can be used because they construct models needed by the user by learning from examples provided by the user [1, 2]. At present, there are three basic directions of research on neural networks [3, 4]:

- classic neural networks (including the multilayer perceptron used in this paper);
- SVM (*Support Vector Machine)* networks,
- neural networks using fuzzy logic (fuzzy neural networks).

In this paper neural networks trained with the backpropagation algorithm method were used for constructing a geometrical displacement model on the basis of geodesic measurements. Geodesic measurements provide a large amount of data about the impact of mining on the rock mass and the land surface, which also makes it easier to explain the process of deformation of engineering structures. Geodesic monitoring, which consists of measurements and their interpretation, makes it possible to draw conclusions about the dynamics of the occurrence of deformations. A typical symptom of deformation are displacements of measurement points located in the research area, caused by changes in the ground-water conditions or by mining [5].

2 Methods

Multilayer feedforward artificial neural networks (multilayer perceptron) can generalize acquired knowledge, i.e. a network that is properly trained on a training data set should generate the expected results after data from the testing set are introduced at the input. Therefore, it is possible to say that a network trained on a set of training patterns transfers its knowledge onto the testing sets as a measure of the ability to generalize. In order to obtain correct generalization, the training process should meet the following conditions [6]:

- the training data should be representative, and their number should be greater than the number of network parameters (weights) that are being determined,
- the network architecture should be as simple as possible, i.e. possibly with the smallest number of weights (an excessive number of neurons in the

* Corresponding author: m.mrowczynska@ib.uz.zgora.pl

hidden layer does not improve the quality of data processing),
- the most effective algorithms should be used in the training process, and they should stop operating when the minimum error is obtained in the testing set.

In a multilayer perceptron network neurons are arranged in layers and only successive layers are connected (Fig.1), for this reason the signal flows in the network only in one direction. The basic problems that should be solved in order to obtain satisfying results while training networks with the error backpropagation method are:
- initial data analysis of the (division of the data set into the training set and the testing set, methods for normalizing the data set),
- network architecture (the number of hidden layers and the number of neurons in these layers), initial selection of network weights, optimum selection of training parameters η and λ, a proper network training process (selection of activation functions, selection of a gradient optimization method, adoption of a criterion for stopping the algorithm).

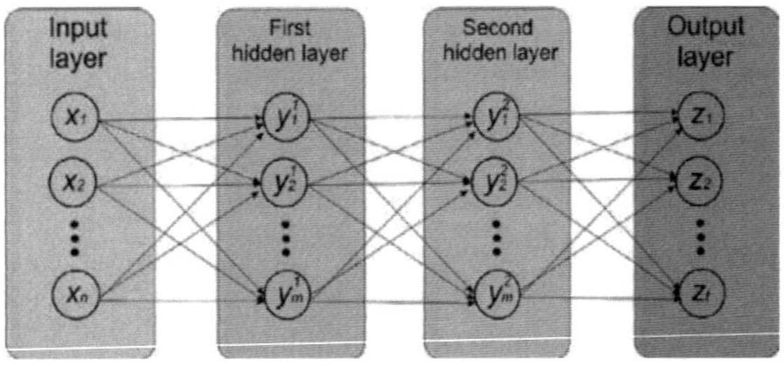

Fig. 1. The structure of a multilayer neural network.

The process of training a neural network is a supervised process (training with a teacher) with the use of gradient optimization methods. It is effected by minimizing the objective function $E(\mathbf{x}, \mathbf{w})$, most frequently defined as

$$E(\mathbf{w},\mathbf{x}) = \sum_{i=1}^{m}(\mathbf{z}(\mathbf{x}_i,\mathbf{w}) - \mathbf{d}_i) \quad (1)$$

where: $\mathbf{w}$ – the matrix of weights, $\mathbf{x}$ – the vector of input signals, $\mathbf{d}$ – the vector of signals expected at the output, z – the output signal obtained as a result of network operation, m – the number of training pairs.

Training a network consists in selecting weights of synaptic connections that ensure that the objective function reaches minimum values. Since it is assumed that the activation function used for training a network is a continuous function (usually bipolar or unipolar), gradient methods can be used for minimizing the objective function.

In gradient methods the training process consists in determining a direction of minimization specified by the vector $\mathbf{p}$, which depends on the values of weights of particular synaptic connections [7]. Therefore, the process of correction of the vector of weights in particular iterations (k – iteration number) is as follows:

$$\mathbf{w}(k+1) = \mathbf{w}(k) + \eta\mathbf{p}(k) \quad (2)$$

where: η is an arbitrary training coefficient adopted from the range (0,1). The method for determining the direction of minimization $\mathbf{p}$ in particular training cycles k for the most frequently used gradient optimization methods is presented in Table 1.

Table 1. The method for determining the direction of minimization.

Gradient optimization method	The minimizing direction $\mathbf{p}(k)$
Steepest descent method	$-\mathbf{g}(\mathbf{w})$
Quasi-Newton algorithm	$-[\mathbf{H}(\mathbf{w})]^{-1}\mathbf{g}(\mathbf{w})$
Conjugate gradient method	$-\mathbf{g}(\mathbf{w})+\beta\mathbf{p}(k-1)$
Levenberg-Marquardt algorithm	$-[\mathbf{G}(\mathbf{w})]^{-1}\mathbf{g}(\mathbf{w})$
Resilient backpropagation algorithm	$-\mathrm{sgn}(\mathbf{g}(\mathbf{w}))$

where:
$\mathbf{g}(\mathbf{w})$ - vector of the gradient in the direction $\mathbf{p}$
$[\mathbf{G}(\mathbf{w})]^{-1}$ - matrix of the approximated Hessian
$[\mathbf{H}(\mathbf{w})]^{-1}$ - broad matrix of Hessian
β - the conjugation coefficient dependent on the value of the gradient in two successive iterations

3 Results

During the process of identification and analysis of the impact of mining on the rock mass and land surface, the results of geodesic measurements are especially important, in particular those that make it possible to determine the vertical displacements of measurement and control points of networks set up in mining areas. Geodesic measurements provide data and information describing the impact of mining on the land surface [8], which makes it much easier to explain the process of deformation of engineering structures located in hazard areas.

In order research the impact of mining on the land surface, geodesic measurements were carried out in the Legnica-Głogów Copper Mining Centre, located in the southern part of the Fore Sudetic Monocline (south-western Poland) [9]. The approximation of land deformation in the Legnica-Głogów Copper Mining Centre was carried out with the use of artificial intelligence and based on the results from four measurement campaigns conducted in the years 1967-2014. In that area of about 75,000 ha the measurement and control network had 137 points that were connected with one another with 164 observations (the observations were carried out with the precise levelling method). A diagram of the structure of the geometric displacement model is presented in Fig. 2.

The approximation of land deformation was carried out with the use of multilayer feedforward neural networks, which were trained with the use of all the gradient optimization methods listed in Table 1 and for different network architecture variants (the number of hidden layers and the number of neurons in the layers). The training set consisted of 70% of all points of the measurement and control network, which had been randomly qualified for this set. The other 30% of points constituted the testing set, which did not participate in the training process. With the correct selection of the structure and parameters of the network, the quality of the operation of the network can be assessed on the basis of the RMSE (the root-mean-square-error) for the testing set. This error is described as follows [10]:

$$RMSE = \sqrt{\frac{1}{m}\sum_{1}^{m}(z_i - d_i)^2} \qquad (3)$$

where m is the number of members in the testing set (or training set), d - the expected value of the output signal, z - the value of the output signal obtained as a result of the operation of the neural network.

The results of the network training and testing process, including the most favourable architecture, are presented in Table 2 and in Figs. 3 and 4. The graphical interpretation of the results of the approximation of the land deformation process is concerned with the most and the least favourable result of the network training process.

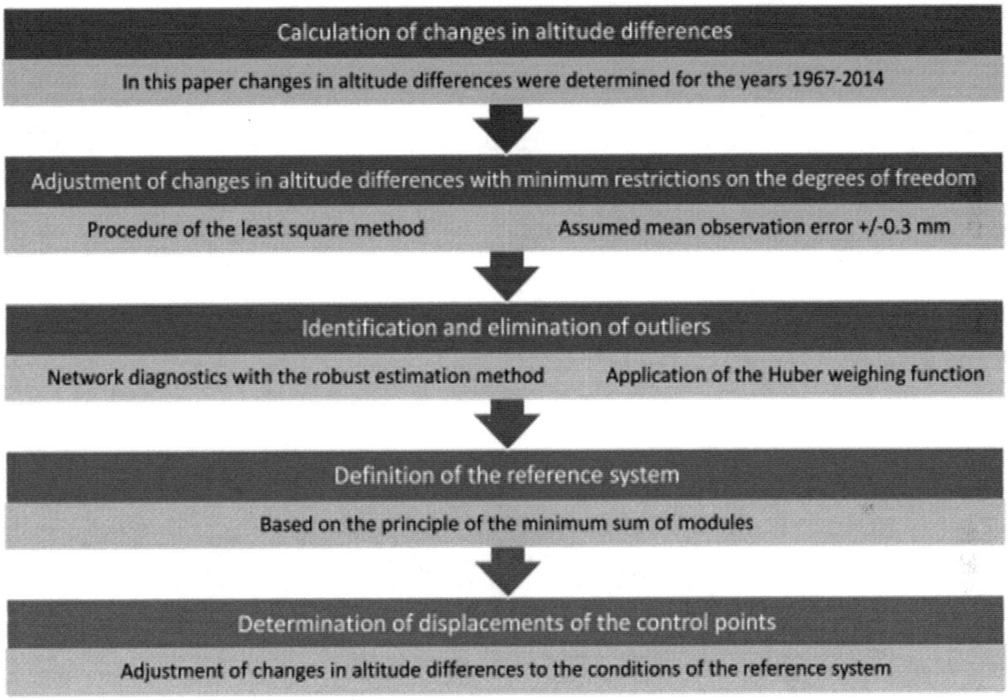

Fig. 2. A diagram of the structure of the geometric displacement model.

Table 2. The results of the neural network training and testing process.

Gradient optimization method	Network architecture	*root-mean-square-error RMSE* [m]	
		Training set	Testing set
Steepest descent method	2_8_3_1	0.29	0.37
Quasi-Newton algorithm	2_7_2_1	0.21	0.23
Conjugate gradient method	2_7_4_1	0.13	0.15
Levenberg-Marquardt algorithm	2_7_3_1	0.17	0.18
Resilient backpropagation algorithm	2_7_1	0.15	0.17
2nd degree polynomial	-	0.33	0.35
Kriging	-	0.12	0.14
Spline	-	0.11	0.13

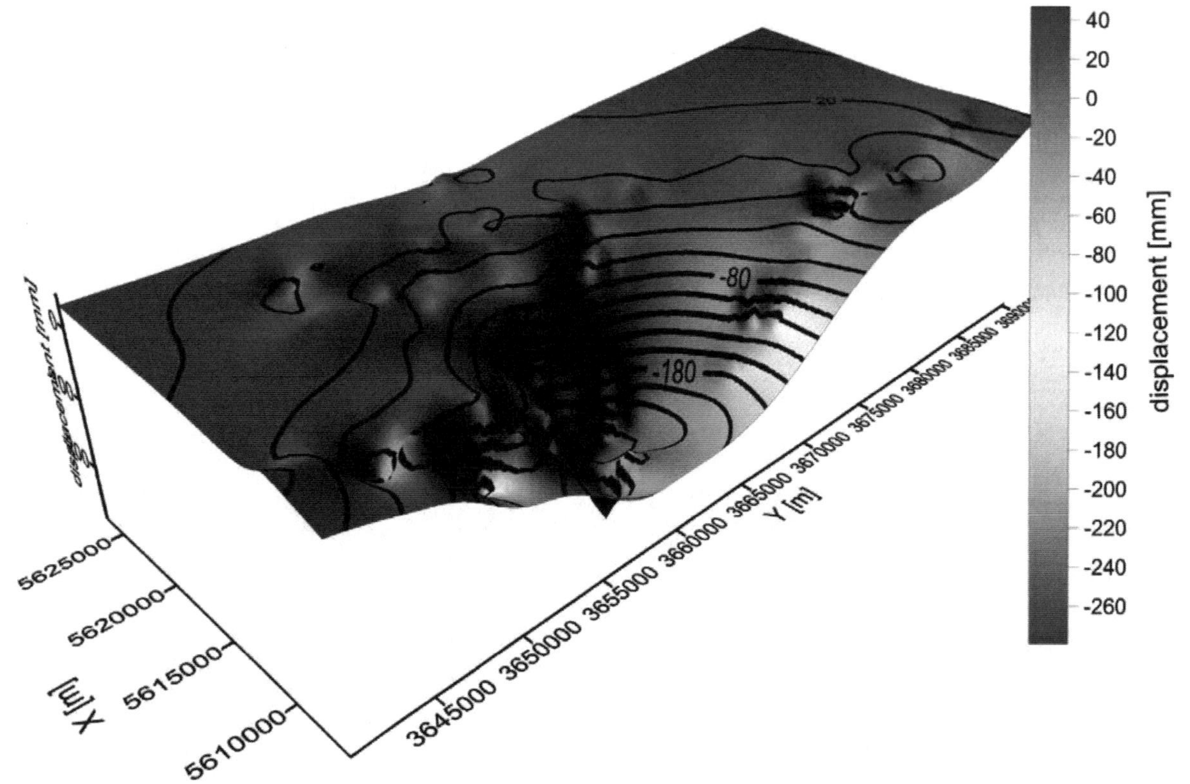

Fig. 3. A geometric model of a land deformation: conjugated gradient method.

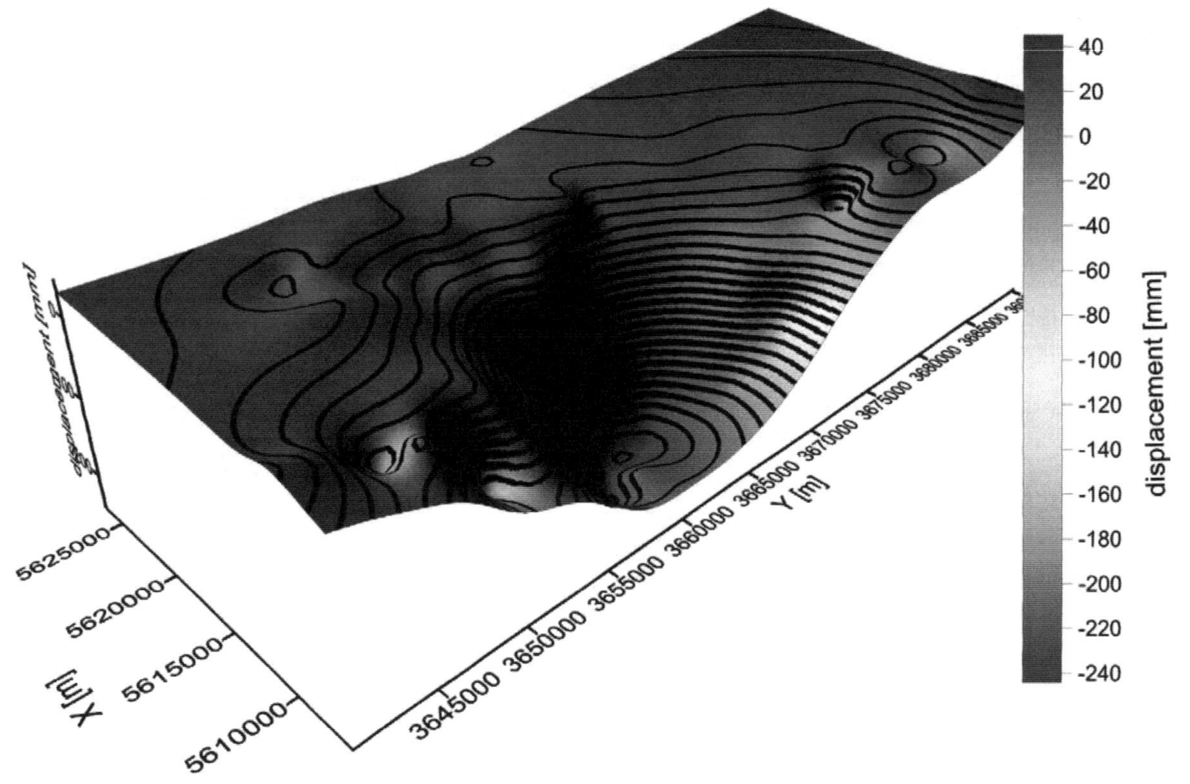

Fig. 4. A geometric model of a land deformation: steepest descent method.

4 Conclusions

The paper presents the possibility of using feedforward neural networks, trained under supervision, for approximation of the process of changes in land surface deformation caused by mining. The neural networks that were used served as a universal approximator, and as a result it was possible to obtain information about changes in land surface occurring in the area under research. The most favourable test results were obtained for a two-layer network with the use of the conjugate gradient method as the training method, and the least favourable results were obtained for the steepest descent method, in which the iteration process very often stopped at local minima. An interesting approach to the problem of training neural networks is the RPROP algorithm, in which the weights are updated dependent on the sign of the gradient components, and its value is disregarded. In the case of this algorithm, the network for which the best result was obtained is a one-layer network, and the results are close to the results obtained by means of the conjugate gradient method.

On the basis of the results presented in the paper it is possible to say that feedforward neural networks work better as instruments for approximating changes in land deformation, and results produced by them are comparable with or slightly better than results obtained with classic methods. An additional advantage of neural networks is the fact that the input set can be updated with additional information such as: the geological structure, ground water level, time interval between geodesic measurements or depth and range of current mining activities, which can reduce the generalization error in the results.

References

1. W. Duch, J. Korbicz, L. Rutkowski, R. Tadeusiewicz, *Sieci neuronowe* (Akademicka Oficyna Wydawnicza Exit, Warszawa, 2000)
2. Ch. Lian, Z. Zeng, W.Yao, H.Tang, Engineering Geology, Volume **186**, 91-99 (2015)
3. J. Barzykowski (Ed.), *Współczesna metrologia. Zagadnienia wybrane* (WNT, Warszawa, 2004)
4. R. Hejmanowski, W.T. Witkowski, Journal of Sustainable Mining, **14**, 101-107 (2015)
5. W.T. Witkowski, Przegląd Górniczy, **2,** 88-94 (2016)
6. M. Mrówczyńska, *Studium nad doborem metod inteligencji numerycznej do rozwiązywania problemów z geodezji inżynieryjnej* (Oficyna Wydawnicza Uniwersytetu Zielonogórskiego, Zielona Góra, 2015)
7. J. Gil, *Przykłady zastosowań sieci neuronowych w geodezji* (Oficyna Wydawnicza Uniwersytetu Zielonogórskiego, Zielona Góra, 2006)
8. R. Kadaj, *Modele, metody i algorytmy obliczeniowe sieci kinematycznych w geodezyjnych pomiar przemieszczeń i dokształceń obiektów* (Wydawnictwo Akademii Rolniczej, Kraków, 1998)
9. M. Mrówczyńska, Przegląd Górniczy, **10**, 37-41 (2015)
10. N. Jankowski, *Ontogeniczne sieci neuronowe. O sieciach zmieniających swoją strukturę* (Akademicka Oficyna Wydawnicza EXIT, Warszawa, 2003)

https://doi.org/10.1051/e3sconf/20185500010

Stability test of TCRP1201+ total station parameters and its setup

Waldemar Odziemczyk[1,*]

[1]Warsaw University of Technology, Faculty of Geodesy and Cartography, 1 Politechniki Sq., 00-661 Warszawa, Poland

Abstract. Monitoring systems utilising automated electronic total stations are an important element of controlling safety levels of numerous engineering structures. In order to ensure appropriate credibility of the obtained data, it is crucial to maintain the stability of both the parameters of the used instrument as well as of its setup. This study presents the results of research on the stability of the Leica TCRP1201+ instrument's basic parameters and three selected tripods manufactured using different materials. In the case of the total station the inspected values included Hz-collimation, V-collimation, and tilt compensation errors. Two of the head tilt components and torsion were examined for each of the tripods. The research was carried out in three-day sessions in conditions of large temperature fluctuations. Graphs presenting the changes in the studied values were supplemented with analysis of correlation between the said changes and the changes in temperature. The conclusions section of the study includes suggestions on how to avoid the negative effects of changes in the geometry of the measuring device.

1 Electronic Total stations in deformation monitoring

Geodetic monitoring of deformations/displacements is an important element of the system ensuring safety of numerous types of engineering structures. It takes advantage of various surveying techniques, such as satellite measurement, photogrammetry, laser scanning, and inclinometric or classic linear and angular measurements. Studies [1-3] contain selected examples of surveying systems utilising the last of the afore-mentioned techniques. A significant advantage of the measurement systems based on the geodetic linear and angular surveys lies in the combination of a relatively high accuracy, its independence from the availability of the satellite signal and the ability to determine absolute displacements referred to an external, from the perspective of the surveyed object, reference system.

Sufficient accuracy of the utilised total station is the first condition for obtaining valuable data of the object state. Papers [4-6] contain information on various methods of investigating the accuracy of instruments. Stability of the total station setup and the points defining the reference system is the prerequisite for the acquisition of adequately accurate and reliable displacement values for the inspected object. In the case of structures subjected to constant monitoring the station setup is usually located on a massive concrete pillar ensuring adequate stability. The situation is somewhat different when carrying out emergency measurements calling for the ability to start the monitoring immediately. As there is not time (and sometimes place) to construct a concrete pillar, the instrument is placed on a tripod. The stability of a total station setup in such a manner depends on a number of factors (type of the tripod, type of the base, changes in the weather conditions) and the possible changes should be taken into consideration when planning the survey and processing the obtained results.

Notwithstanding the foregoing, one should take into consideration the occurrence of changes in the parameters of the instrument itself. In particular it may concern the change in the orientation of line of sight in regard to the telescope which manifests itself as Hz- and V-collimation changes and zero point of the electronic vial used for compensation of the influence of tilt of the instrument's vertical axis.

This study is a continuation of research described in [7] and presents the results of an experiment aimed at researching the changes of the basic parameters of the measuring position and the instrument in the conditions of a several-day-long test survey. Leica TCRP1201+ total station and three types of tripods – aluminium, composite and wooden – were used in the experiment.

Changes in the following parameters were analysed within the framework of this research:

1. For the instrument
   - Hz- and V-collimation,
   - zero point of the electronic vial (tilt compensation system),
2. For each of the tripods
   - Instrument's tilt
   - Changes in the orientation of the horizontal circle

The test site was selected so that the measurements coincided with largest possible changes in the temperature.

* Corresponding author: waldemar.odziemczyk@pw.edu.pl

2 Measurement tests

Measurement test were carried out inside the skylight located in the great hall of the Warsaw University of Technology. The skylight's structure consists of a steel frame and two shells – internal and external. The external shell is made of glass whereas the internal one is made of plastic. On sunny days the enclosed space between the shells causes significant temperature fluctuations. In such conditions changes in the instrument's and the measuring position's parameters would be larger and easier to notice. On the other hand, the limited air exchange inside the skylight causes the changes in the relative humidity to be closely correlated to the changes in temperature.

Both the construction of the skylight and the performance of the surveying instrument were the subject of an earlier research, the results of which were described in [7-10].

Automated Leica TCRP1201+ total station was used in order to carry out the measurement tests. The improved, as compared to the previous versions, ATR (automatic target recognition) system of this instrument was described in [11]. Its software is based on Leica TPS1200 operation system and it allows for control over the work of the total station via GSI interface. This solution was used in TC-calc software working on PC (Windows) in order to carry out the measurement tests. Detailed information on the TC-calc software can be found in the author's earlier papers [12] and [13]. The ability to read the instrument's current tilt value turned out to be a particularly important feature of the TCRP1201+ total station.

The stability of the total station's setup also depends on the quality of the tripod. It is particularly important in the case of instruments equipped with servomotors. The effects of dynamic loads occurring at the beginning and the end of rotation of the instrument were the subject of a study by C. Depenthal [14].

During the test the total station was successively placed on three tripods: aluminium (Leica), composite (Nedo) and wooden (Leica). All of them were standard tripods. It was ensured that all the screws were properly tightened and that all clearances were eliminated on all of the interconnecting parts of each of the tripods.

The measurement test entailed automatic measurement of two points 1 and 2 in complete series every 30 min. The instruments setup was located on a concrete pillar supporting the structure of the skylight. The locations of points 1 and 2 were selected in such a way, that the lines of sight to them formed a right angle (Fig. 1). The measurement took 64 to 72 hours, which allowed for the observation of several cycles of daily temperature fluctuations.

Horizontal and vertical angles were registered during the measurements (Hz, V) and the tilt components of the vertical axis (Tx, Ty) were read.

Test measurements were performed in summer. Three measurement sessions were carried out in total. Detailed information on each of the sessions are juxtaposed in Table 1.

Horizontal and vertical angles were registered during the measurements (Hz, V) and the tilt components of the vertical axis (Tx, Ty) were read from the instrument's compensation system.

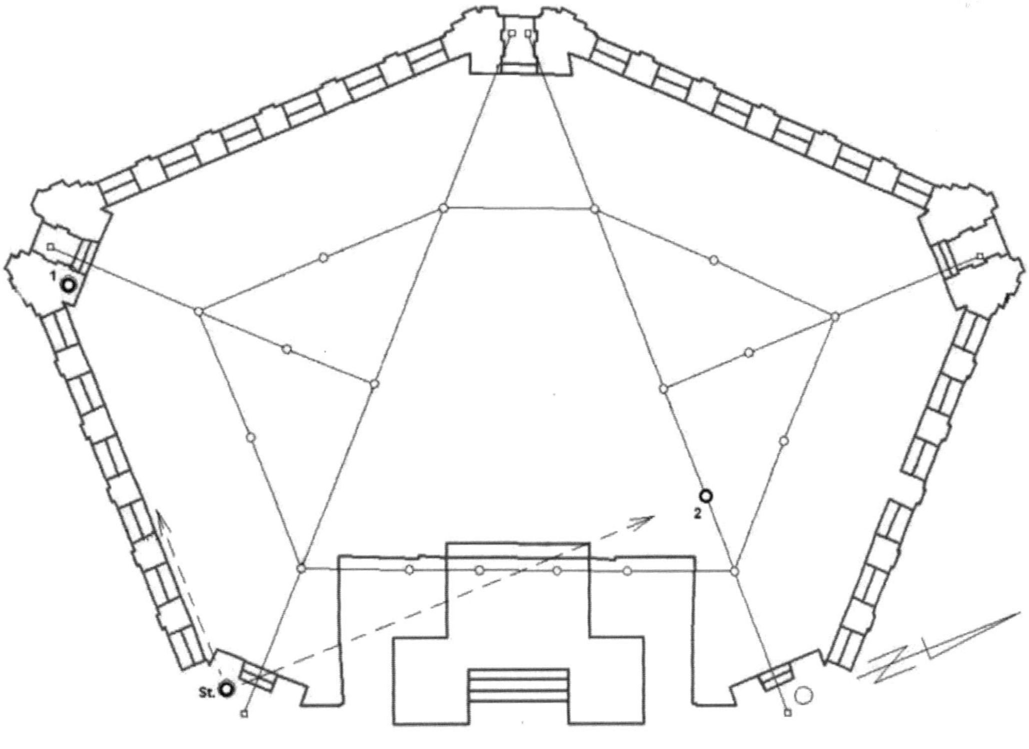

Fig. 1. Location of the station and control points during test measurements.

Table 1. Measurement tests.

No.	Type of tripod	Measurement time [h]	T min. [°C]	T max. [°C]	Amplitude T [°C]
1	aluminium	64	20.2	42.8	22.6
2	composite	72	18.4	40.1	21.7
3	wooden	72	23.2	46.0	22.8

3 Processing of the measurement results

The obtained results of the measurements were used to track the changes in the analysed parameters for the instrument and the setup. The first step of result processing involved averaging of the horizontal *Hz* and vertical *V* angles obtained for each of the series (*i*).

$$Hz_i = \frac{Hz_{i(KL)} + Hz_{i(KP)} - 200^g}{2} \quad (1)$$

$$V_i = \frac{V_{i(KL)} - V_{i(KP)} + 400^g}{2} \quad (2)$$

The values for *Hz*-collimation *k* and *V*-collimation *M* for *i* series were determined using the following equations:

$$k_i = \frac{Hz_{i(KL)} - Hz_{i(KP)} \pm 200^g}{2} \quad (3)$$

$$M_i = \frac{V_{i(KL)} + V_{i(KP)} - 400^g}{2} \quad (4)$$

In each of the measurement series the values of collimation errors and zero of the vertical circle were determined twice (independently for each of the observed points).

For each point, except of horizontal and vertical angles, indications of the instrument's tilt compensation systems were registered in the form of two components: longitudinal (*TL*) and transverse (*TC*) referred to the instrument's system. The tilt determined in the first half-series was noted as:

- TL_{1A} – longitudinal tilt while measuring point 1
- TC_{1A} – transverse tilt while measuring point 1
- TL_{2A} – longitudinal tilt while measuring point 2
- TC_{2A} – transverse tilt while measuring point 2

The tilt determined in the second half-series was noted analogically TL_{1B}, TC_{1B}, TL_{2B}, TC_{2B}.

The way of utilising measured instrument tilts to determine orientation of total station vertical axis is described in [15]. Because the directions to points 1 and 2 form a right angle (Fig. 1), the tilt components of the instrument towards each of the points can be determined 4 times. Thus, the tilt of the instrument's vertical axis towards point 1 (*Tx*) will be equal to:

$$Tx = TL_{1A} = -TC_{2A} = -TL_{1B} = TC_{2B} \quad (5)$$

Analogically, the tilt in the direction of point 2 (*Ty*) will be equal to:

$$Ty = TC_{1A} = TL_{2A} = -TC_{1B} = -TL_{2B} \quad (6)$$

Mean value of all four determinations will be assumed as the final value:

$$Tx = \frac{TL_{1A} - TC_{2A} - TL_{1B} + TC_{2B}}{4} \quad (7)$$

$$Tx = \frac{TC_{1A} + TL_{2A} - TC_{1B} - TL_{2B}}{4} \quad (8)$$

It has to be noted, that the determined values of the tilts, due to the use of both the locations of the alidade, are free from the influence of the fixed error (zero point) of the compensation system.

In order to determine the zero points of the compensation system one uses the readings of the appropriate tilts (longitudinal and transverse) to the same point in different placement of the alidade (half-series).

$$TL_0 = \frac{TL_{1A} + TL_{1B}}{2} = \frac{TL_{2A} + TL_{2B}}{2} \quad (9)$$

$$TC_0 = \frac{TC_{1A} + TC_{1B}}{2} = \frac{TC_{2A} + TC_{2B}}{2} \quad (10)$$

The final values for a measurement series took the form of a mean value of all the corresponding calculations.

$$TL_0 = \frac{TL_{1A} + TL_{1B} + TL_{2A} + TL_{2B}}{4} \quad (11)$$

$$TC_0 = \frac{TC_{1A} + TC_{1B} + TC_{2A} + TC_{2B}}{4} \quad (12)$$

The change in the averaged horizontal angle from the measurement point to point 1 was assumed as the measure of the changes in the orientation of the tripod. The point was located on a concrete prop supporting the skylight and can be treated as fixed.

4 Study of changes in the instrument's parameters

As previously mentioned in chapter 1, the instrument's performance is described by four parameters. Two of them relate to the position of the line of sight (electronic) in relation to the telescope's casing. The remaining two are the zeroes of the instrument's tilt compensation system. Due to the fact that those parameters are not dependent on the used tripod, each of the measurement sessions can be used for independent determination of the listed parameters.

Because each of the sessions yielded concordant results, the author decided to present the result of a single measurement session no. 1 carried out using the aluminium tripod.

As it can be observed in Fig. 2 and 3, despite large temperature fluctuations the changes in the Hz- and V-collimation are relatively small. It stands in contrast with the analogical results obtained for TDA5005 total station [9]. The course of changes in the line of sight indicates a small dependency of the changes in the measured value on the temperature. However, the changes are minimal in relation to the accuracy of the measurement, so no unequivocal conclusion can be drawn.

The analysis of the changes in the zeroes of the instrument's tilt compensation system yielded more interesting results. The graph presented in Fig. 4 shows that the instrument's tilt compensation system is characterised by different behaviour for different components. As far as the zero for the transverse component (*TC*) is concerned, it is virtually resistant to temperature changes. However, in the case of longitudinal component (*TL*), a clearly visible fluctuations tied to the temperature changes with the amplitude equal to approx. 13^{cc} can be observed. Correlation between zero position for the *TL* component and the temperature is confirmed by Pearson correlation coefficient (Fig. 5). Minor decentration of the graph indicates that the changes in the parameters of the compensation system occur with a slight delay (approx. 2.5 – 3 h) in relation to the temperature change.

Further research is required in order to explain the occurrence of this phenomenon.

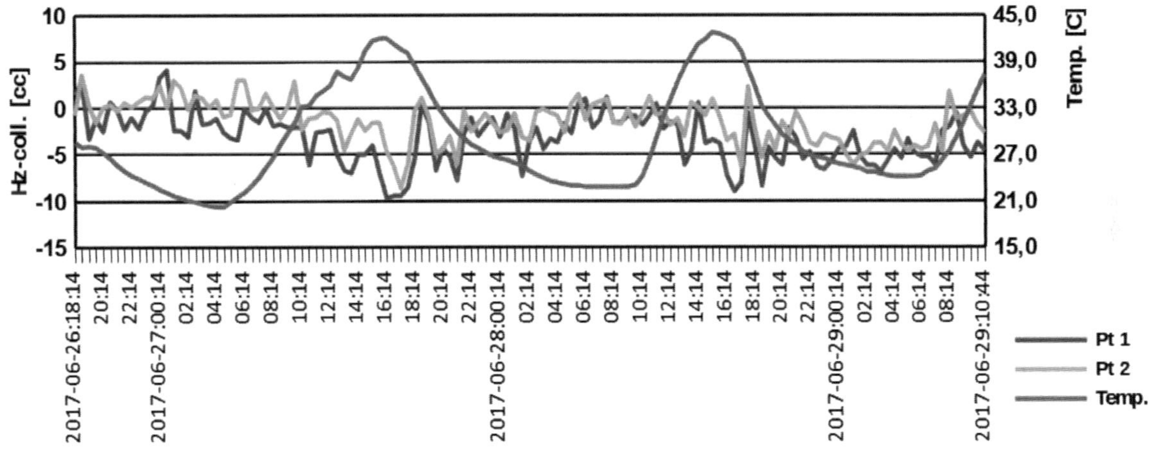

Fig. 2. Changes of Hz-collimation.

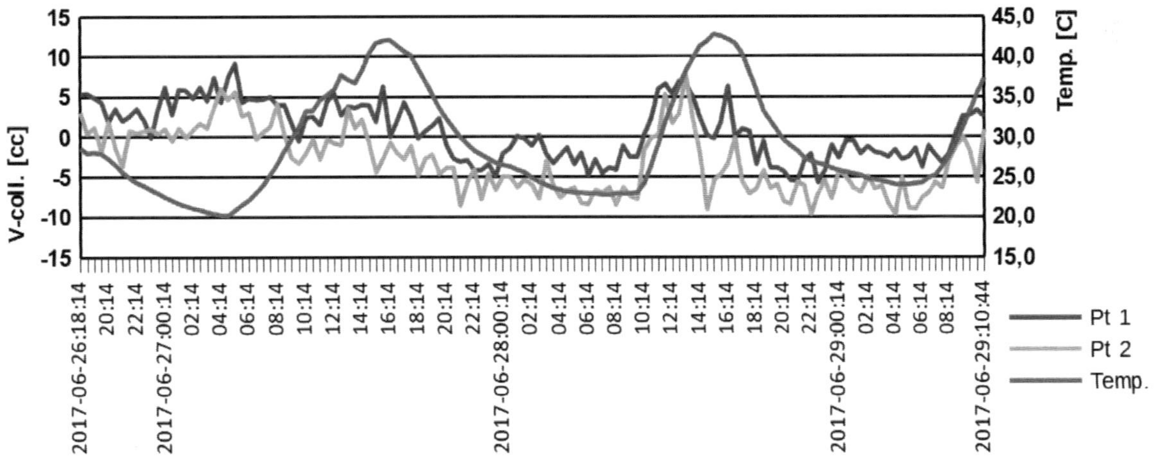

Fig. 3. Changes of V-collimation.

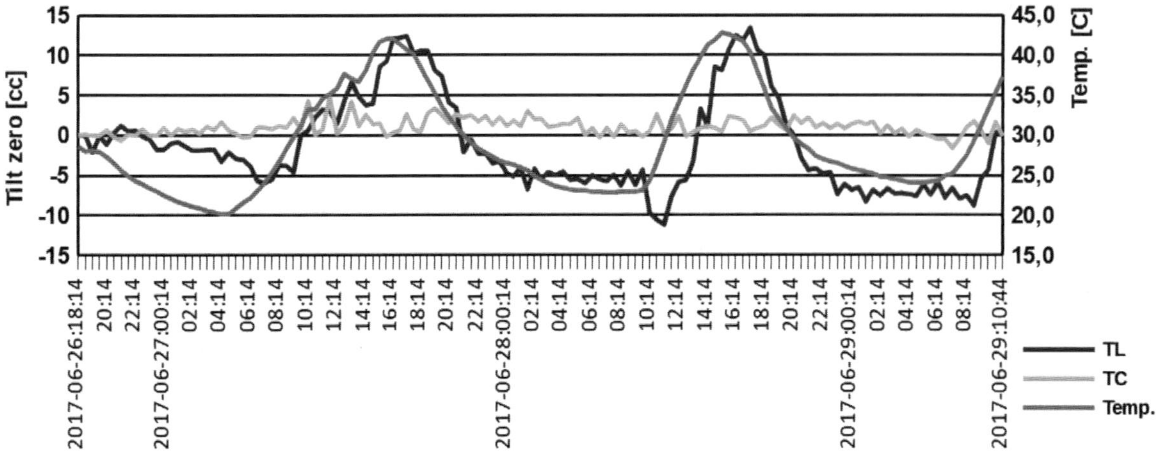

Fig. 4. Changes of tilt zero point of compensation system.

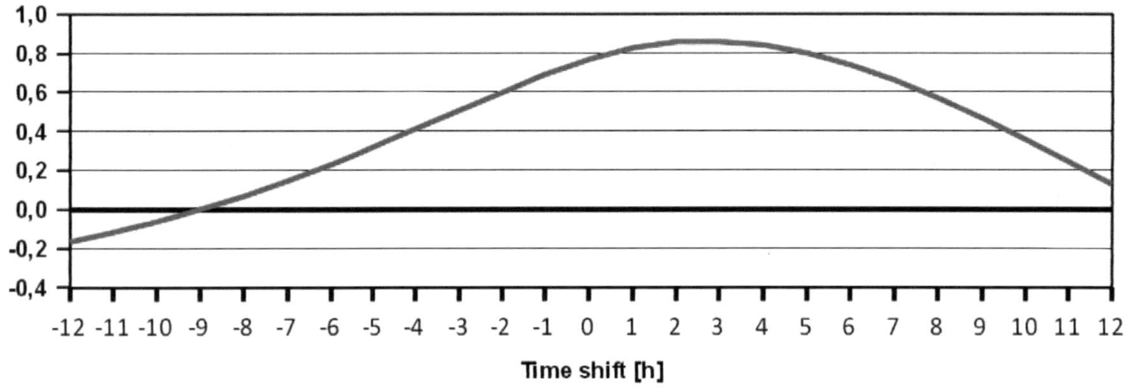

Fig. 5. Correlation between TL_0 and temperature (Pearson coefficient).

5 Study of tripod stability

The scope of this research also included stability checks for the selected types of tripods. Using the function of GeoCom interface of the TCRP1201+ instrument the changes in the total station's vertical axis tilt (Tx, Ty) identified as changes in the orientation of the plane of the tripod's head were scrutinized. Change in the horizontal angle measured towards the assumed fixed point 1 was the torsion of the tripod. Changes in the tilt of the vertical axis of the instrument for each of the tripods are presented in Figs. 6, 7 and 8.

In order to facilitate the interpretation of the presented data red colour was used to track temperature fluctuations.

Figures 9, 10 and 11 present the changes in horizontal orientation for each of the successive measurement sessions.

https://doi.org/10.1051/e3sconf/20185500010

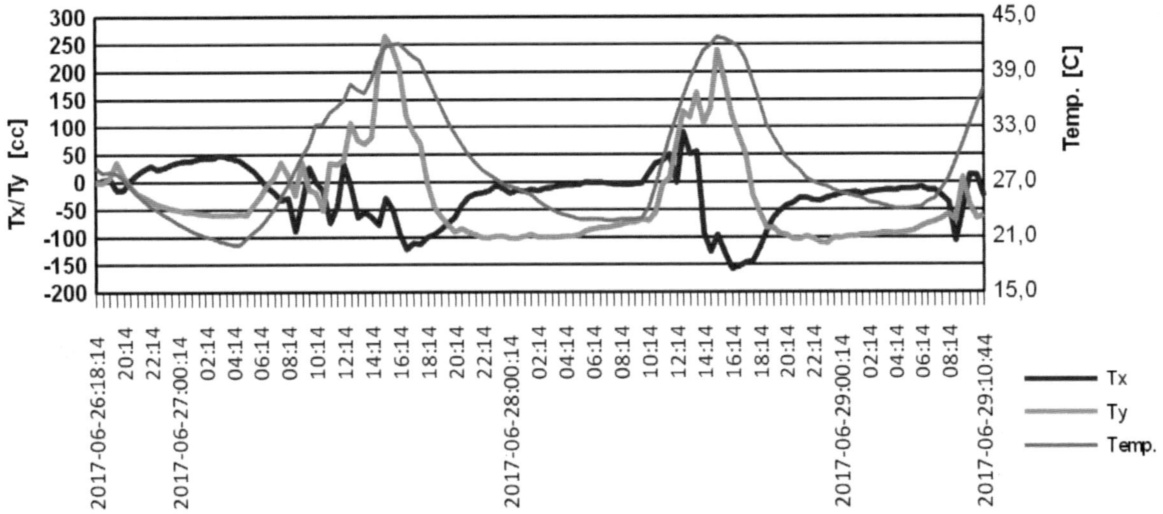

Fig. 6. Changes of instrument tilts for aluminium stand.

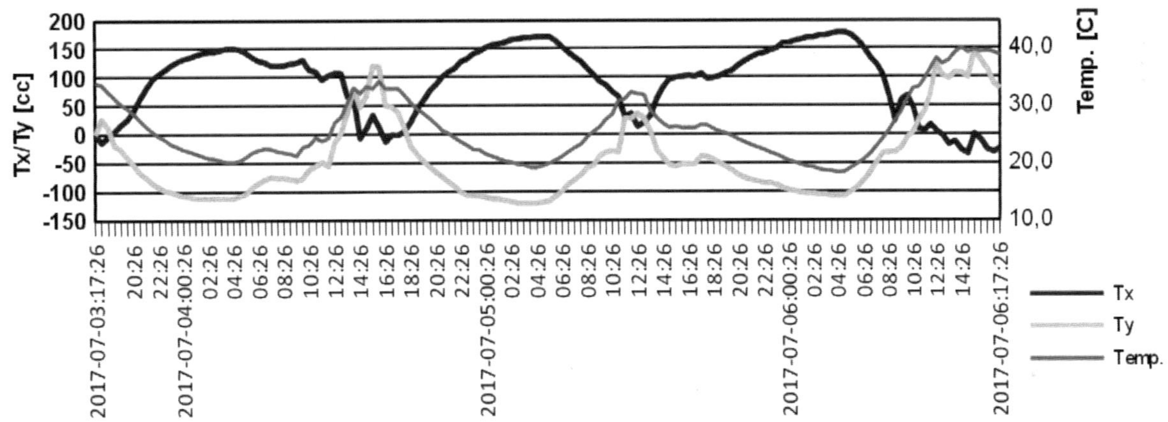

Fig. 7. Changes of instrument tilts for composite stand.

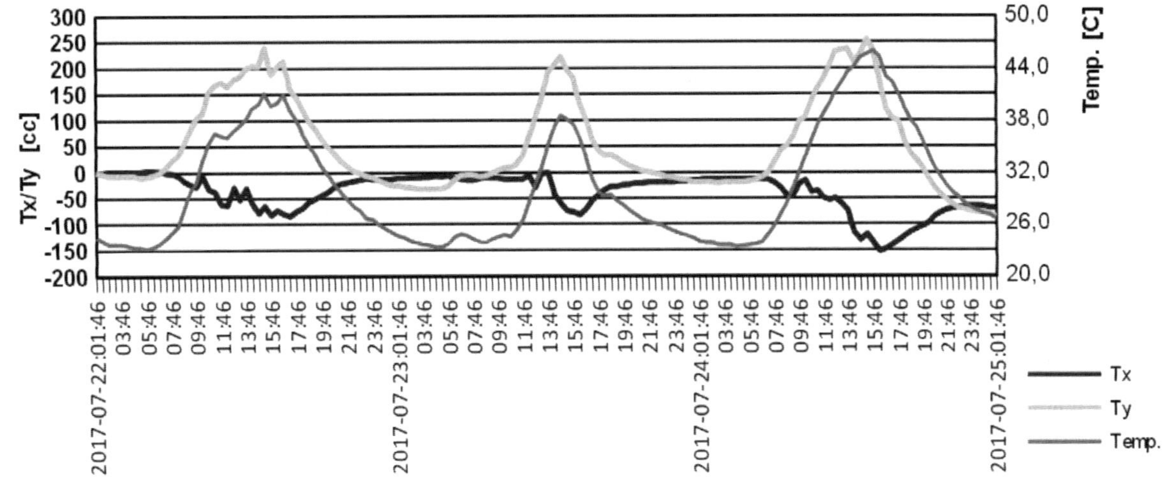

Fig. 8. Changes of instrument tilts for wooden stand.

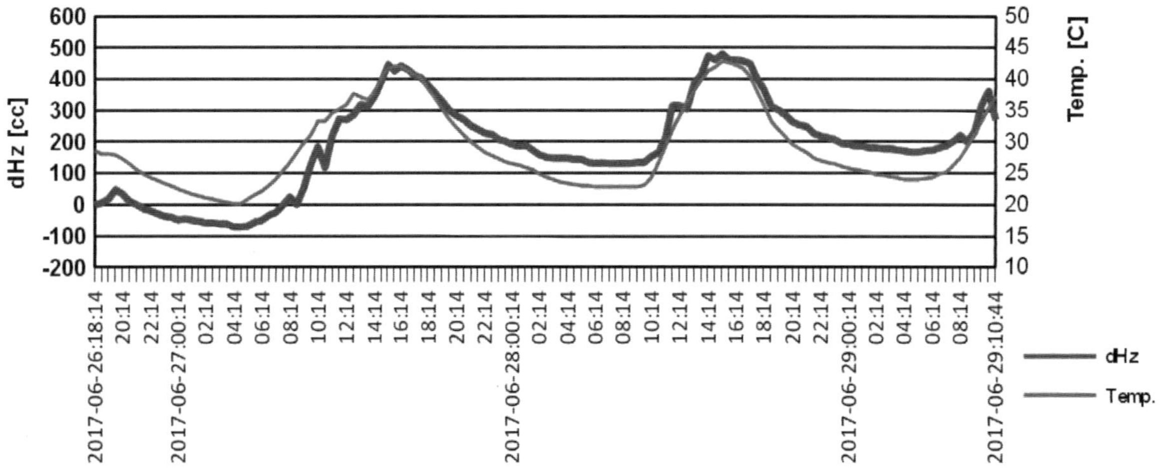

Fig. 9. Changes of instrument orientation for aluminium stand.

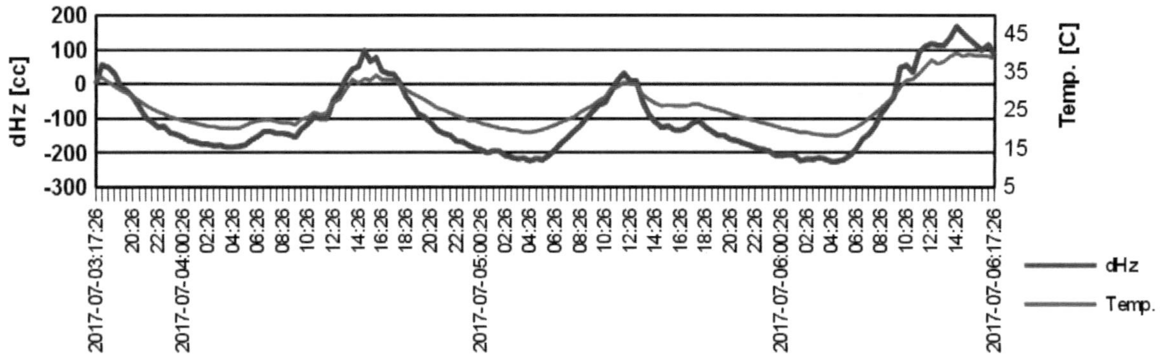

Fig. 10. Changes of instrument orientation for composite stand.

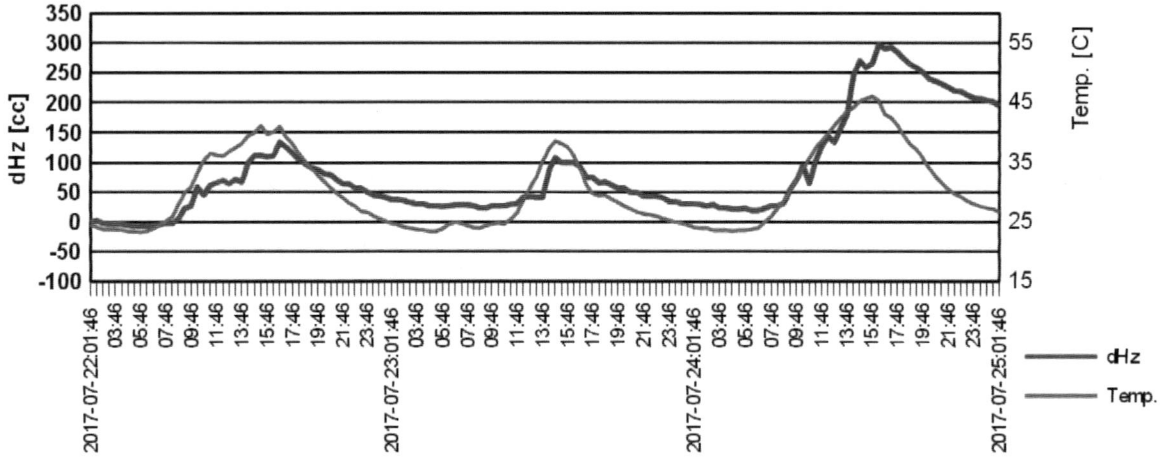

Fig. 11. Changes of instrument orientation for wooden stand.

It can be easily noticed, that both the changes in the tilt of the instrument's vertical axis are closely associated with the temperature changes. This fact is confirmed by the values of the Pearson correlation coefficient presented in Figs. 12, 13 and 14.

The tilt values are rather high, and in the case of the aluminium and wooden tripods they reached 250^{cc}. Slightly smaller tilt values were noted for the composite tripod. Moreover, it is noticeable, that regardless of the type of the tripod the tilt signs were the same, which may be connected to the direct solar operation (the tripods were not shielded from the sunlight).

Change in the orientation for each of the tripods was also clearly connected to the temperature fluctuations,

https://doi.org/10.1051/e3sconf/20185500010

which was confirmed by the analyses of the correlation coefficient presented in Fig. 15. By analyzing the orientation change for each of the tripods it can be noticed that apart from repeatable changes, a certain steady, approximately uniform (Fig. 9) or variable in time (Fig. 11), trend is observable. Consequently, after the temperature returned to the initial value the instrument positioned on a tripod was orientated differently.

The carried out tests show that the smallest changes in geometry occurred for the composite tripod. Furthermore, those changes are, to the greatest extent, tied to the temperature (Figs. 13, 15). The wooden tripod retained stability for most of the time. However, during the last day of the tests in the period characterized by the highest temperatures an increased change in its geometry occurred for reasons unknown, which influenced the results of statistical analysis and lowered the Pearson correlation coefficient. Analysing Figs. 12 and 14 (aluminium and wooden tripods) it can be noticed, that *Tx* and *Ty* changes are similarly shifted in relation to changes of temperature. In author's opinion influence of direct sunlight on the tripods geometry can be taken into consideration as an explanation of that similarity.

When summarizing the obtained results it has to be noted, that the character of changes in tripod geometry depends not only on the type of material used for its manufacture, but also it is, to a large extent, an individual characteristic of a particular unit. Consequently, the conclusions of studies concerning tripods can be only of general nature and can only relate to orders of magnitude and the character of the said changes.

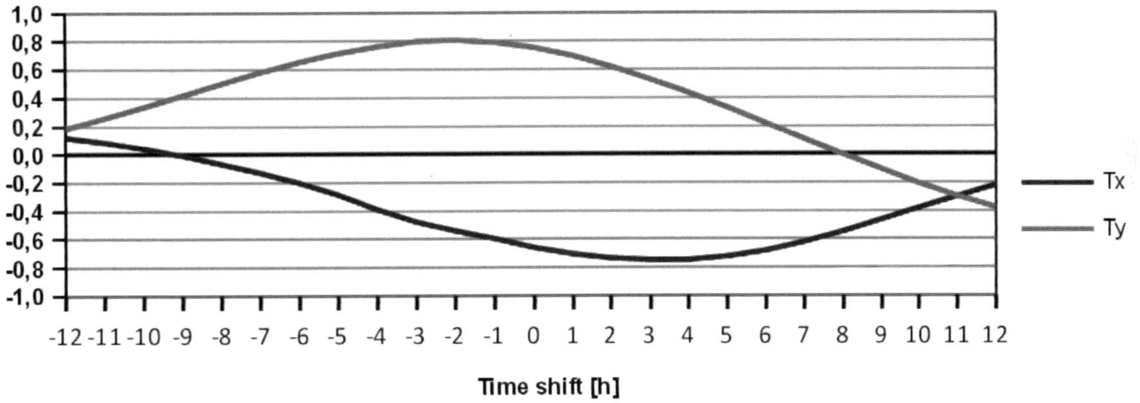

Fig. 12. Correlation between instrument tilts and temperature for aluminium stand (Pearson coefficient).

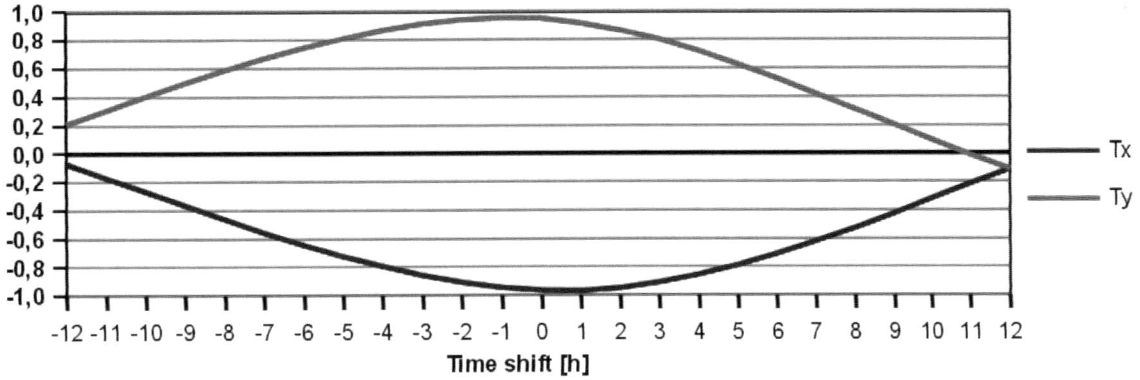

Fig. 13. Correlation between instrument tilts and temperature for composite stand (Pearson coefficient).

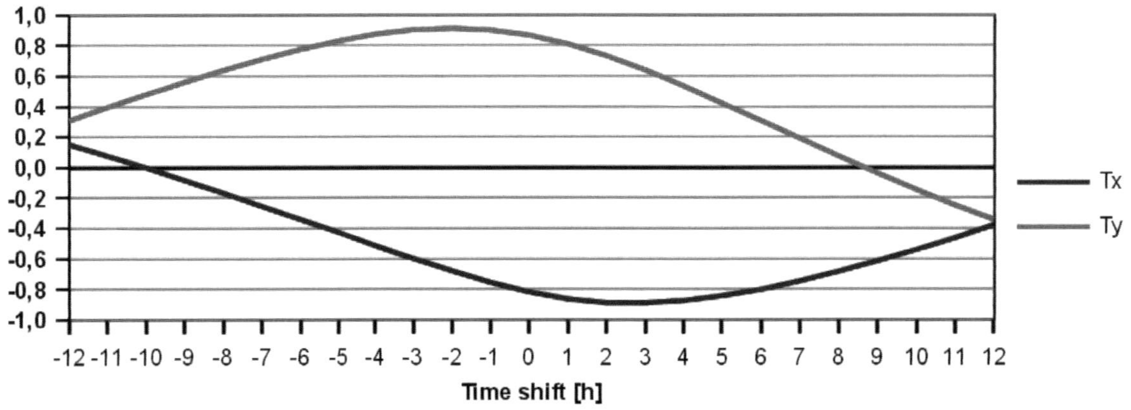

Fig. 14. Correlation between instrument tilts and temperature for wooden stand (Pearson coefficient).

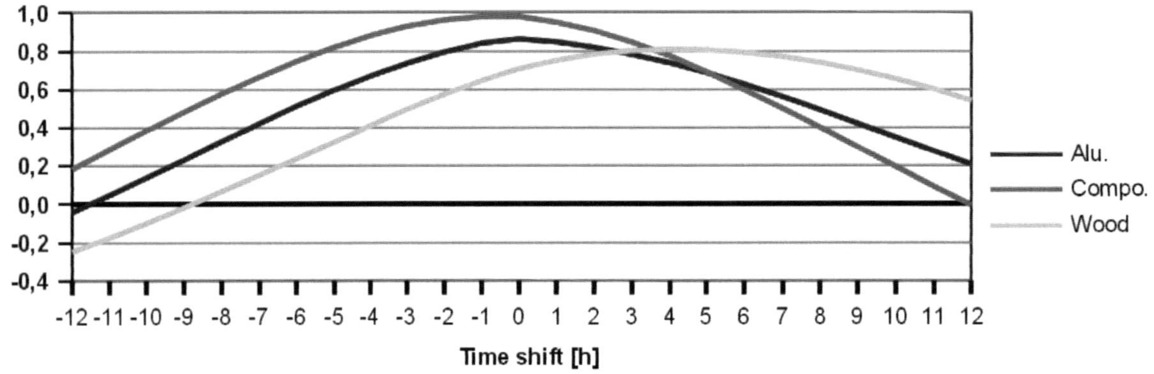

Fig. 15. Correlation between changes of orientation and temperature (Pearson coefficient).

6 Summary

One can draw the following conclusions on the basis of the obtained measurement results:

The studied instrument (TCRP1201+) turned out to be relatively highly resistant to fluctuations of external temperature, which makes measurements in one face a viable option. This possibility is further supported by the fact, that in the real field conditions the temperature fluctuations will rarely reach such high values as those observed during the test measurements. It should be noted however, that the one face measurement should be treated as the last resort and be used only in situations where there is no possibility to carry out full measurement series.

The measurement in two faces is also supported by the results of study on the behaviour of the total station's vertical axis tilt compensation system. As proven by the tests, the system is partially (one component) susceptible to temperature fluctuations. Measurement in two faces makes it possible to eliminate errors caused by the tilt compensation system.

The study on tripod stability revealed that the changes in the geometry of standard tripods used in geodetic surveys are significant and in the case of an extended measurement have to be taken into consideration.

The performed measurement tests were accompanied by large temperature fluctuations. That translated into significant geometry value changes in the inspected tripods, which was confirmed by the analyses of Pearson's correlation coefficient. The composite tripod turned out to be the most stable among all the tested variants. It was the result of moderate geometry value changes as well as their highest correlation with temperature fluctuations. It does not, however, change the conclusion, that in the case of all of the tripods those changes were significant in relation to the angle measurement accuracy required in the control measurements of engineering structures.

Although the changes in the tilt of the tripod's head translating into changes in the tilt of the instrument's vertical axis usually do not exceed the range of its compensation system, the torsion of the tripod resulting in the change of the horizontal orientation ought to be taken into account during the measurement design stage. Periodic measurements towards fixed point placed outside the object's zone of influence can be used for that purpose.

Regardless of the above, in the case of long-term measurements one should aim to use heavy tripods with reinforced construction which guarantee higher level of stability for the instruments setup.

https://doi.org/10.1051/e3sconf/20185500010

References

1. V. Michel, T. Person, M. Kasser, Surveying in Industry and Construction FIG Working Week, **TS22.1** (2003)
2. A. Berberan, M. Machado, S. Batista, Surv. Rev., **39**(305), 203-211 (2007)
3. W. Stempfhuber, Bautechnik, **89**(11), 794-800 (2012)
4. T. Owerko, M. Strach, Rep. on Geod., **87**(2)/2009, 291-299 (2009)
5. D. S. Paraforos, M. Reutemann, G. Sharipov, R. Werner, H. Griepentrog, Computers and Electronics in Agroculture **136**/2017, 166-175 (2017)
6. J. Vaněček, M. Štroner, *Acta Polytechnica* **56**(6), 478-491 (2016)
7. M.Woźniak, W. Odziemczyk, Rep. on Geod. and Geoinf., **104**/2017, 79-90 (2017)
8. M. Woźniak, W. Odziemczyk, S. Jastrzębski, J. Wojciechowski, Rep. on Geod., **1**(84), 105-110 (2008)
9. M. Woźniak, W. Odziemczyk, Rep. on Geod. **1**(85), 105-110 (2009)
10. W. Odziemczyk, Rep. on Geod. and Geoinf., **97**/2014, 35-46 (2014)
11. H. Kirschner, W. Stempfhuber, (Proceedings of the 1st International Conference on Machine Control & Guidance, 51-60, 2008)
12. M. Woźniak, W. Odziemczyk, *Wykorzystanie metod geodezyjnych w ocenie stanu geometrycznego budowli* (Publishing House of Silesian University of Technology, 173-180, 2008, in Polish)
13. W. Odziemczyk, *Inżynieryjne Zastosowania Geodezji* (Publishing House of Poznan University of Technology, 2012, in Polish)
14. C. Depenthal, Allgem. Verm.-nachrichten **6/2004**, 227-233 (2004)
15. P. Raby, U.S. Patent No. 6,243,658 (2001)

Proposition of determination of displacements using the TDRA 6000 laser station

Jacek Sztubecki[1,*], *Maria* Mrówczyńska[2], and *Adam* Bujarkiewicz[1]

[1]University of Science and Technology, Faculty of Civil and Environmental Engineering and Architecture, 7 prof. S. Kaliski St., 85-796 Bydgoszcz, Poland
[2]University of Zielona Góra, Institute of Building Engineering, Department of Land and Environment Engineering, 9 Licealna St., 65-417 Zielona Góra, Poland

Abstract. With modern measurement technologies, it is possible to achieve higher precision and solve complex measurement tasks. The tested device (TDRA 6000 laser station) features a direct drive technology ensuring a 3D point accuracy of 0.25 mm. TDRA 6000 combined with proper software is a perfect tool to monitor engineering objects. The article presents the measurement technique and the method of calculation of 3D displacements of control and measurement networks using the TDRA 6000 laser station from Leica. The values presented herein derive from periodic measurements carried out in control points stabilised at a ground the above-ground crossing between buildings located on the premises of the University of Science and Technology campus in Bydgoszcz.

1 Introduction

Geodetic monitoring with measurements and their interpretation makes it possible to specify conclusions on the dynamics of changes in objects [1, 2]. The basic symptom of adverse phenomena occurring on buildings includes vertical displacements of control and measurement network points [3]. Contemporary measurement methods can be applied to complete tasks which could not be fulfilled with the traditional methods, while ensuring the required accuracy. Furthermore, measurements conducted with such techniques are less labour intensive and less time consuming [4].

Displacements and deformations of engineering structures can be determined by using modern laser devices. They allow to get better accuracy and give the ability to solve complex measurement tasks. Today, these technologies are often used in metrological industrial measurements [4]. They are characterized by high precision, reaching +/- 15 μm + 6 μm/m when measuring with a retroreflector and operating range up to 120 m [5]. These parameters are obtained for measurements of average size of structures.

The article presents an analysis of results of vertical and horizontal displacement measurements carried out with the TDRA 6000 laser station from Leica. The station features a direct drive technology ensuring a 3D point accuracy of 0.25 mm. The producer's inspection certificate in accordance with DIN 55350-18-4.2.2 specifies the following accuracy for this device: maximum deviation of distance measurement of +/- 0.2 mm (standard deviation of single distance measurement σ=0.1 mm) and standard deviation of angle measurement (Hz and V) σ=0.15 mgon [6]. The device combined with proper software is a perfect tool to monitor engineering objects.

Methods applied in measurements may constitute an alternative solution to measurements conducted with classic methods. The article also presents the way of stabilisation of measurement points. With an analysis of measurement data, it was possible to make conclusions on the measurement technique and precision.

2 Tested objects

Displacement determination measurements with the use of the TDRA 6000 laser station were conducted on two control networks established on the buildings of the University of Science and Technology in Bydgoszcz.

The existing benchmarks stabilised on the external walls of the 3.1 building were adapted to the first network (Fig. 1).

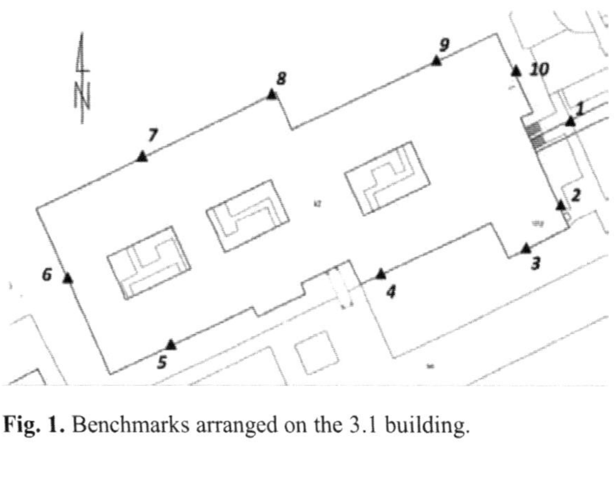

Fig. 1. Benchmarks arranged on the 3.1 building.

* Corresponding author: jaceksz@utp.edu.pl

Steel nuts used to stabilise a 1.5" red-ring-reflector were permanently fixed on the benchmarks (Photo 1).

Photo. 1. 1.5" red-ring-reflector stabilisation.

In order to determine vertical displacements, the difference in elevation between the respective points was measured. The measurements were conducted at equal target distances not exceeding 30 m. Each measurement on every stand was carried out at least two times, while changing the laser station height and making sure that the deviation from elevation differences was permissible.

The second network was established inside above-ground crossing between buildings which connects the Auditorium Novum building with the 2.1 building (Photo 2).

Above-ground crossing between buildings is of a reinforced concrete design supported on 12 columns. The network points were stabilised with washers (Photo 3) fixed to the floor of above-ground crossing and the adjacent buildings. The measurement network arrangement layout is shown in Fig. 2.

The measurements were conducted on 7 free stands (Fig. 2) while observing the position of the 1.5" red-ring-reflector and recording 3D coordinates of the control points. Each measurement stand was used twice, moving slightly the measuring device.

Photo. 2. The above-ground crossing between buildings.

Photo. 3. Stabilisation of the control points on the floor inside the above-ground crossing between buildings.

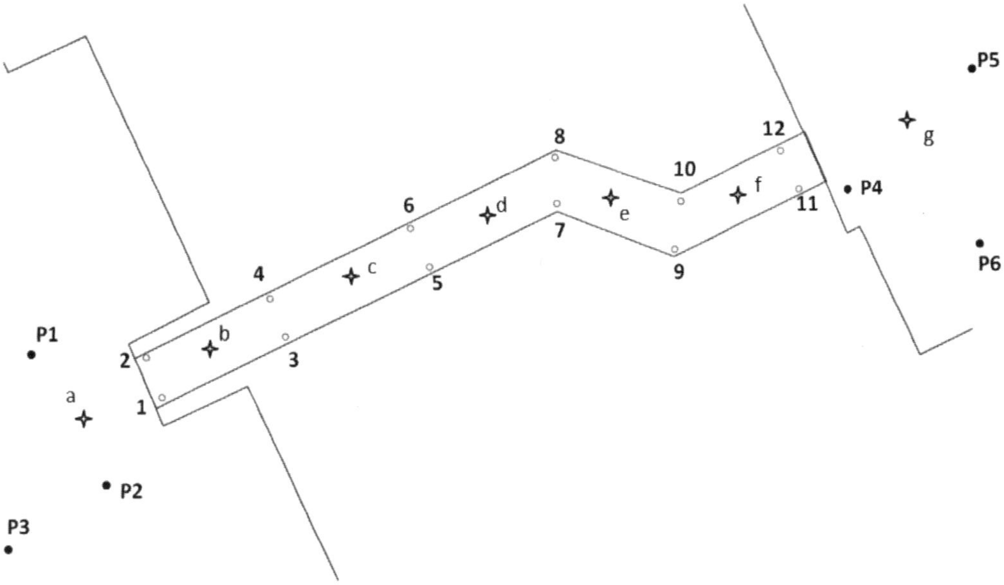

Fig. 2. Control point arrangement layout (P1÷P6 – reference points, 1÷12 – control points, a÷g – instrument stand).

3 Test results

Four measurement series were conducted for the first network. An initial measurement check was performed following each series by summarising and adding up elevation differences in a circuit. The differences in the height and value concerning not-closing of the mesh of the measured network was presented in Table 1.

Due to the lack of points showing stability in the height range in the vicinity of the examined object, in the process of leveling the network the relative vertical displacements were measured.

Relative displacement values and their errors are presented in Table 2. It can be noticed that errors of determined displacements do not exceed 0.20 mm, which is within the generally accepted limit of vertical displacement measurement accuracy.

In the adjustment process, the error values of measured height differences were also determined. The maximum values of these errors were:
- 1 measurement – 2 measurement – 0.03 mm,
- 1 measurement – 3 measurement – 0.08 mm,
- 1 measurement – 4 measurement – 0.06 mm.

The procedure for the second network involved two measurements of 3D point coordinates on each of 7 stands (a-g). Only the nearest control points were observed from each station: for the outermost stands (a, g) – 5 points, for the remaining stands – 4 points.

Two measurement series were conducted for which calculations of spatial control point displacements were made at two stages.

At the first stage, differences in elevation between the network points were calculated and then subjected to rigorous adjustment. Results obtained at this stage and vertical displacement errors are shown in Table 3.

Table. 1. Results of network measurement checks for individual measurement series.

Elevation difference [mm]					
Point	*measurements*				*Point*
	1	*2*	*3*	*4*	
1					*1*
	-18.0	-18.1	-18.3	-18.3	
2					*2*
	377.2	377.2	377.0	376.8	
3					*3*
	1223.6	1224.2	1224.0	1223.7	
4					*4*
	5.1	5.1	5.0	3.9	
5					*5*
	-295.7	-295.7	-295.5	-295.4	
6					*6*
	253.0	253.0	253.5	253.6	
7					*7*
	-178.2	-178.3	-178.4	-178.3	
8					*8*
	62.8	62.8	63.2	63.8	
9					*9*
	-321.2	-321.2	-321.2	-321.0	
10					*10*
	-1108.6	-1109.1	-1108.9	-1108.5	
1					*1*
Total	**0.0**	**-0.1**	**0.4**	**0.3**	

Table. 2. Values of displacements and error of displacements.

Point	Displacements and error of displacements [mm]						Point
	1 measurement – 2 measurement		1 measurement – 3 measurement		1 measurement – 4 measurement		
	dZ	mdZ	dZ	mdZ	dZ	mdZ	
1	0.08	0.04	0.58	0.16	0.76	0.16	*1*
2	-0.01	0.03	0.24	0.12	0.43	0.12	*2*
4	0.61	0.03	0.36	0.12	0.07	0.12	*4*
5	0.62	0.04	0.22	0.16	-1.16	0.16	*5*
6	0.63	0.05	0.38	0.18	-0.89	0.18	*6*
7	0.64	0.05	0.84	0.20	-0.32	0.20	*7*
8	0.55	0.04	0.60	0.18	-0.45	0.18	*8*
9	0.56	0.05	0.96	0.20	0.52	0.20	*9*
10	0.57	0.05	0.92	0.18	0.69	0.18	*10*

Table. 3. Results of network measurement checks for individual measurement series.

Point	dZ [mm]	mdZ [mm]
1	0.15	0.07
2	0.08	0.07
3	-0.05	0.07
4	-0.09	0.07
5	-0.13	0.08
6	-0.22	0.08
7	-0.01	0.08
8	0.17	0.08
9	-0.07	0.07
10	0.06	0.07
11	0.03	0.07
12	-0.18	0.07

The second stage involved determination of horizontal displacements of the established network. Calculations were made with flat X, Y coordinates of the control points in the following sequence:

1. an isometric transformation of points obtained from two positions of the laser station (separately for each measurement station) was performed for each free stand;
2. an isometric transformation combining (integrating) the free stands into one unit was performed for each measurement series;
3. a Helmert transformation was performed on 6 adjustment points for both measurement series – the purpose of the transformation was to present control point coordinates in a uniform system of coordinates;
4. a transformation of control point coordinates was performed for both measurement series up to the system of an axis parallel to the section connecting points 1 and 7;
5. dX and dY displacements were calculated between the measurement series as differences of X, Y coordinates.

As a result of the calculations, horizontal displacements of the control points were obtained. These are presented in Table 4.

An analysis of the accuracy of determined displacements was based on a transformation error. The value of this error was defined for points 1 and 3 of the calculation order.

Errors were determined for isometric transformation calculations carried out in accordance with point 1. The value of these errors is presented in Table 5.

An error of 0.49 mm was determined for Helmert transformation calculations carried out in accordance with point 3.

Table. 4. Results of horizontal displacements.

Point	dX [mm]	dY [mm]
1	0.45	0.94
2	0.44	0.85
3	0.16	1.18
4	0.22	1.13
5	-0.35	0.77
6	-0.35	0.66
7	-0.87	0.19
8	-0.84	0.11
9	-1.57	0.19
10	-1.69	0.14
11	-1.89	0.28
12	-2.09	0.13

Table. 5. Results of horizontal displacements.

Stand	Transformation error [mm]	
	***0 measurement* 2017.07.24**	***1 measurement* 2017.09.09**
a	0.11	0.08
b	0.23	0.16
c	0.20	0.21
d	0.10	0.24
e	0.15	0.08
f	0.12	0.05
g	0.28	0.19

4 Summary and conclusions

The article presents possible applications of the TDRA 6000 laser station from Leica designed to determine vertical and horizontal displacements of engineering buildings.

Vertical displacement measurements covered a network of benchmarks located outside the 3.1 building which formed a closed circuit. The results confirmed the expected measurement accuracy and indicated that an average elevation difference error did not exceed 0.13 mm already at the network control stage when checking the circuit closure. Further analysis involved determination of observation errors and uncertainties based on rigorous adjustment of the measured network. It indicated that average errors of measured elevation differences did not exceed 0.12 mm and the error of determined displacements was not greater than 0.20 mm.

Analysed measurement results of vertical displacements of the control points located inside the above-ground crossing between buildings show their high accuracy of 0.08 mm. Horizontal displacements of this network were determined with isometric and Helmert transformations. Both isometric transformation

https://doi.org/10.1051/e3sconf/20185500011

errors for the free stands (max 0.28 mm) and the Helmert transformation error between the measurement series (0.49 mm) indicate high accuracy of the applied measurement method.

Measurement accuracy achieved when carrying out the test confirms that the employed methods constitute an alternative to classic displacement determination methods.

References

1. T. Lazzarini, *Geodetic measurements of deformations and their application in construction,* (PPWK, Warsaw, 1961, in Polish)
2. B. Wolski, *Metrological monitoring of geotechnical objects* (Publishing House of Krakow University of Technology, Krakow, 2006, in Polish)
3. R. Kadaj, *Models, methods and algorithms for kinematic networks in geodetic measurements of displacements and deformation of objects* (Publishing House of Krakow University, Krakow, 1998, in Polish)
4. A. Bujarkiewicz., J. Sztubecki, M. Sztubecka, Materiały Budowlane **503**(7), 50-51, 2014
5. G.J. Wojcik, S.A. Lakanen, Proceedings of the 7th International Workshops on Accelerator Alignment, SPring-8, Japan, 2002
6. Leica Geosystems, http://metrology.leica-geosystems.com/, access10 July, 2017

Examination of vertical displacements at non-permanently fixed points

Robert Gradka[1,*]

[1]Wrocław University of Science and Technology, Faculty of Geoengineering, Mining and Geology, 27 Wyb. Wyspiańskiego St., 50-370 Wrocław, Poland

Abstract. The aim of this paper was to analyze vertical displacements in four different test areas, with various types of surfaces, without permanently stabilized measuring: points and control network, using the same measurement method. The first stage involved measuring selected test areas with a number of measuring points that reliably reflect the surface and to construct the DEM. The second stage involves acquiring of a grid of control points, evenly distributed in the test area, used to assess the accuracy of the DEM. The last stage involved the analysis of the acquired results, including the application of an original program to calculate vertical displacements. In order to conduct the analysis the "blind test" was introduced, which allowed to perform formal analyses, i.e. verification of the accuracy of models constructed based on direct measurements. In this program (based on the IDW interpolation method), the study of the dependence of introducing fixed points (given measurement error), to measured survey points was carried out. It was possible to determine vertical displacements between measurements in each area, calculate the accuracy of the mapping of the terrain surface, fit the DEM into the actual terrain, as well as obtain the parameters of the approximate planes.

1 Introduction

Deformations of terrain occur mainly due to the changes of load on the ground, in the level of groundwater, as well as any land sliding. Also, the progressing investments (deep excavations or other earthworks) may pose a threat to neighboring buildings and may be the reason for monitoring the displacements of land surface and structural monitoring of buildings [1, 2]. Especially in the areas of mining activities and other hazardous areas, measurements of deformations and displacements in terrain include important land surveyors' work [3, 4]. In such areas, measurements of displacements and deformations provide information on the hazard level to building security and can prevent the occurrence of a construction disaster. In addition, the displacement measurement results are also used to verify assumed displacement values [1, 5]. The choice of measuring method and the terrain deformation study method are determined by a number of factors, including [6]:

- the desired accuracy of evaluation of point displacements, dependent on the types of analyzed structures,
- the density and size of the control grid,
- the type of displacements determined (relative, absolute, vertical, horizontal, 3D),
- the rate of changes occurring.

Determining terrain deformation, such as vertical and horizontal displacements, requires highly accurate land surveying measurements. In classical land surveying, determination of vertical displacements commonly employs the method of precise leveling, while in the case of horizontal displacements, angular and linear measurements in control grids [7-10]. The author has decided, however, that the determination of displacements in the terrain will involve geometric leveling, ant not precise leveling, since the measurement will be performed on random points, placed on the ground, not permanently fixed thereon.

In the literature, you can find many items on surveying vertical displacements, however, mainly on fixed points in the terrain or directly on surveyed structures. The author attempted to examine vertical displacements, both on measurement points and on the geodetic control network which have not been permanently stabilized in the terrain, between the first and the second measurement.

2 Characteristics of test areas and measurement methods

Four areas located in the city of Krakow were measured twice (the first measurements took place in 2016, the second in 2017). Two of them were areas with low vegetation (grass), while the third is the old airport (concrete), and the last one - the area of one of the city parks (mixed: asphalt and grass). The shape, layout and location of the measurement dot grid is presented in the Fig. 1 (example of area no. 1), in the plane geodetic

* Corresponding author: robert.gradka@pwr.edu.pl

 https://doi.org/10.1051/e3sconf/20185500012

coordinate system – "PL-2000". Each of the research area was approximately 95 m long and 65 m wide.

The study areas were selected in such a manner so that the points of the geodetic network were as close as possible to each other. The points of the measuring network were stabilized only temporarily in the terrain and located in a different place, for each of the measurements. Each of the survey point was marked out according to the assumed measurement grid (11x11 points), consisting of a total of 121 points (Fig. 1), in each of the areas (total: 968 of survey points for both measurements). The points were distributed evenly in relation to one another, according to the GRID [11, 12], measured (using a TOPCON GPT-3000 total station) and marked for the duration of the measurements with a peg, pin or a nail (depending on the test area). Each of the point (survey and network) was also measured using the GNSS satellite measurement technique – the Trimble R8 receiver, and finally measured using the Leica Sprinter 150M code leveler to obtain the heights needed for analysis. The measured points also served to build and evaluate the GRID DEM accuracy [11-13].

To perform the repeated measurements, all points were marked on each object using GNSS technique (without stabilization), and then aluminum leveling staff was placed near the original point and re-measured with a code leveler. Due to the fact that the author did not want to faithfully reproduce the points from the first measurement, but only to be in the approximate position in relation to the points from the initial measurement, hence it was reasonable to use the GNSS receiver to determine the points. The accuracy of the determination of points from the first measurement is not accidental. The leveling staff has the spacing of ca. 60 mm at the base, therefore, after determination of points with GNSS with the accuracy of ±30 mm, the probability of measuring a point at the very same (initial) spot or in its very near proximity is high. The re-measurement was to show differences in the height of the points, from the initial and subsequent measurements, the same terrain relief (the same range of the grid) and to investigate the displacements of the objects (if any occur) using non-fixed points and only reproduced by the GNSS technique.

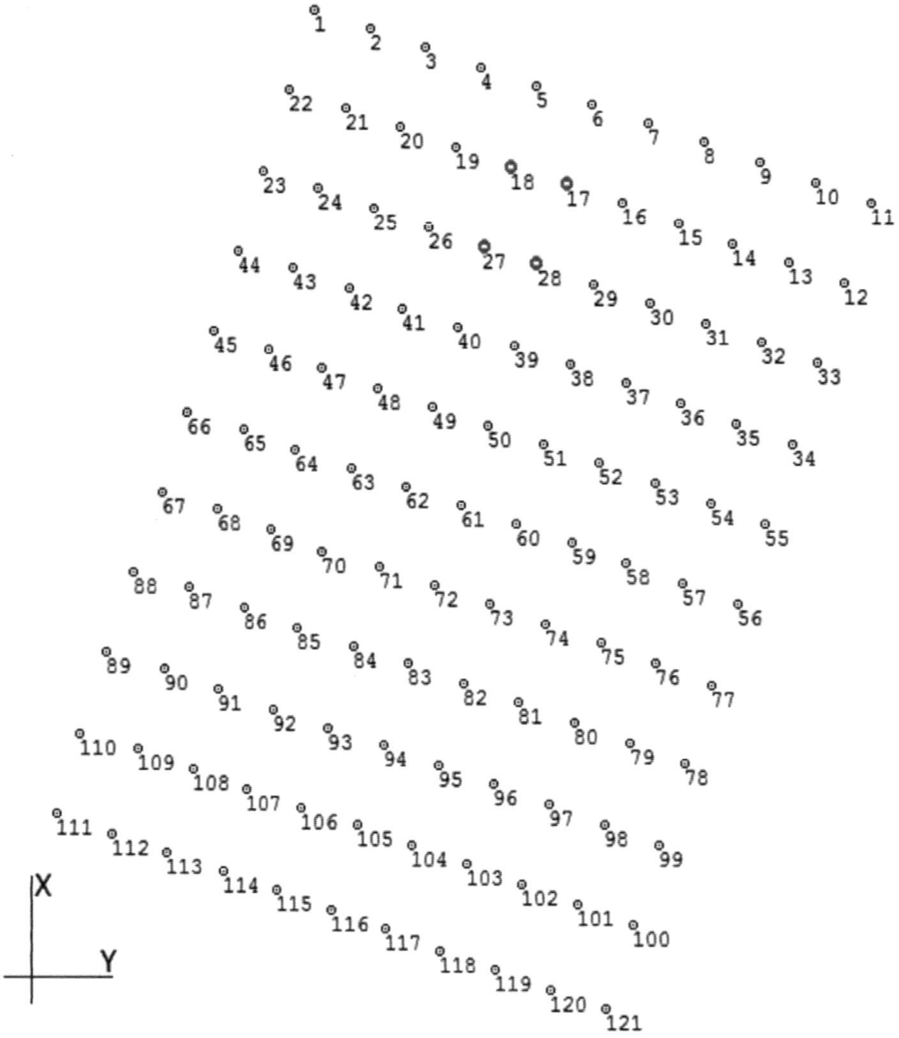

Fig. 1. The shape, layout and location of the points of the measurement grid (example area 1).

3 The analysis of vertical displacements and plane parameters

Knowing that in a small area mass displacements can be approximated with any plane [7, 14] the question arises: is it possible to obtain pre-set values of plane parameters, with different combinations of non-fixed points (in initial and secondary measurement) and theoretically fixed points (adopted for the time of calculation as a permanent but not fixed in the field in reality)? The "blind test", carried out by the author, was an experiment used to determine vertical displacement between measurements in a given research areas, depending on the number of fixed points entered (with a given measurement error) to the survey points.

It was decided to analyze the data obtained from direct measurements in the original program, based on the data interpolation method – IDW (Inverse Distance Weighting) [14-16]. The IDW method was chosen for the analysis due to the easiness to program the algorithm. For each measured areas, several additional conditions and parameters were introduced:

- all coordinates are in a local system – untransformed to the "PL-2000" – coordinate system,
- GRIDs were generated in the shape and size of the rectangles consisting of the next four measurement points, nearest to each other (Fig. 1 – red dots / points no.: 17, 18, 27, 29) – coincide with a grid of measurement points in a local system,
- the number of fixed points is: 0-5 – randomly selected among measured points,
- measurement error for a fixed point is ±2 mm – determined by the author,
- average error for measuring the height of the point is ±30 mm,
- nearest point searching radius is 10m,
- value of function for coordinates is a plane with 0 parameters of the general equation of the plane,
- number of repetitions for calculations per each area and each fixed point is 50,
- **ex, ey, H0** - vectors of changes in the position of the calculated plane from the second measurement, relative to the plane obtained from the first measurement – plane "0").

The program works by choosing a predetermined number of random points from a given area, with a pre-established height error based on a given number measurement and repetitions of calculations. The program was created in the C++ environment.

For each area, data needed to be prepared for the calculations in the form of text files (Tables 1-3). For the first measurement, the file (Table 1) contains as follows: coordinates (local) X, Y, height above sea level H and average error for measuring the height of the point mH [mm].

Table 1. File fragment with data for the calculations (first measurement).

X	Y	H	mH
30.570	245.714	202.789	30
33.668	240.376	202.797	30
36.774	234.971	202.837	30
39.840	229.613	202.875	30
50.919	228.852	202.592	30
47.834	234.240	202.585	30
44.727	239.603	202.604	30

The text file for the secondary measurement (Table 2) - in consecutive lines, the coordinates (local) X, Y, height H, the average error for measuring the height of the point mH [mm] and the value of the function for the coordinate data Hfunk – the plane with the 0 parameters.

Table 2. File fragment with data for the calculations (secondary measurement).

X	Y	H	mH	Hfunk
30.570	245.714	202.809	30	0
33.668	240.376	202.814	30	0
36.774	234.971	202.850	30	0
39.840	229.613	202.884	30	0
50.919	228.852	202.596	30	0
47.834	234.240	202.584	30	0
44.727	239.603	202.611	30	0

GRID network file (Table 3) - in consecutive lines, the coordinates (local), X, Y, the function value for Hfunk coordinates data.

Table 3. GRID file fragment with data for the calculations.

X	Y	Hfunk
111	223	0
114	223	0
117	223	0
120	223	0
123	223	0
12	226	0
15	226	0
18	226	0

Control file (Table 4) of the calculating program contains (in lines):

1. Interpolation distance for a given point [m] (the maximum range in which the points are taken to interpolation),

2. safety parameter [m] (if the measuring point lies exactly where computation point, it would be an error, because there would be a division by zero),
3. measurement error for a fixed (adopted as a permanent) point [mm] – determined by the author,
4. number of generated fixed (adopted as a permanent) points,
5. number of repetitions for calculations,
6. file localization and name series 1,
7. file localization and name series 2,
8. grid file localization and name,
9. results file localization and name.

Table 4. Control file fragment with data for the calculations.

Line	Attribute
1	10.0
2	0.1
3	2.0
4	0
5	50
6	D:\s1.txt
7	D:\s2.txt
8	D:\grid.txt
9	D:\result.txt

The result of the calculations is the result file (Table 5), which contains, in the line for each given calculation result, consecutively:

- number of the calculation,
- mean difference of height between series 1 and series 2,
- root mean square error (RMSE) - the root of the sum of error squares model estimation,
- ***ex*** (vector component along the X axis) parameter of the plane function - difference between the value of the "0" function and a plane approximated by calculating the proprietary program,
- ***ey*** (vector component along the Y axis) parameter of the plane function - difference between the value of the "0" function and a plane approximated by calculating the proprietary program,
- ***H0*** (vector component along the Z axis) parameter of the plane function - difference between the value of the "0" function and the values of functions interpolated by the program; the actual displacement relative to the "0" plane and the 0 height.

4 Received results

For each area, the results of the calculations for a given number of fixed (adopted as a permanent) points with 50 repetitions of calculations are presented in tables (6-9). The tables contain data from the result files, i.e: the average difference between series [m], root mean square error (RMSE) [m], three plane parameters (***ex***, ***ey***, ***H0***), as well as the number of fixed points [items] and calculated minimum and maximum differences between series [m].

Table 5. Fragment of the calculation result file.

No.	Difference	RMSE	ex	ey	H0
1	0.0030	0.0103	0.003040	-0.001237	0.00121
2	0.0031	0.0107	0.003023	-0.001281	0.00475
3	0.0026	0.0100	0.003049	-0.001289	0.00474
50	0.0020	0.0101	0.003060	-0.001283	0.00337

Table 6. Compilation of calculations from the result file for area 1 in the local system.

Number of fixed points	0	1	2	3	4	5
Min. difference between series [m]	-0.005	-0.0054	-0.0051	-0.0051	-0.0052	-0.0052
Max. difference between series [m]	-0.005	-0.0046	-0.0034	-0.003	-0.0031	-0.0029
Average difference between series [m]	-0.005	-0.0049	-0.0046	-0.0045	-0.0044	-0.0043
RMSE [m]	±0.0098	±0.0096	±0.0103	±0.0102	±0.0104	±0.0106
Parameter ex	0.000029	0.000029	0.000026	0.000025	0.000025	0.000025
Parameter ey	-0.00005	-0.00005	-0.000045	-0.000048	-0.000041	-0.000045
Parameter H0	-0.00149	-0.00135	-0.0015	-0.00112	-0.00155	-0.00105

https://doi.org/10.1051/e3sconf/20185500012

When analyzing Table 6, it can be seen that the mean square error of the difference between the two series of measurements increases with the number of fixed points and is about ±1 cm. The average difference between series has a negative value and is about -5 mm with an increasing trend. If you take into account the parameters of the "0" plane, ex and ey have a decreasing tendency, while H0 is unevenly distributed. In the even-points, it increases, and in odd ones it decreases. Area no. 1 was characterized by slight changes in the shape of the terrain, but with the large height difference, hence the results are ambiguous.

It can be read from table 7 that the average difference between the series of measurements is in plus and it is within the range of ±1 cm. The mean difference error is ±1.5 cm. Both attributes show a decreasing trend with a growing number of fixed points. The mean error is greater than in the case of area 1 (Table 6), despite smaller land sliding. The ex and ey parameter distributions are differentiated and do not show a clear trend – they rise and drop at different rates. At the same time, the ***H0*** parameter decreases with the increase of fixed points.

The compilation of calculated values for area 3 is shown in Table 8. Based on this, it can be stated that the mean square difference error is 3 times greater in the case of area 2 (Table 7) and 2 times greater than in area 1 (Table 6). The error is ±5.3 mm. The average difference between the series is ca. +14mm. As for the plane parameters, all three tend to increase in value (***ex*** and ***ey***), while the ***H0*** parameter decreases with the increase of fixed points randomly selected for the calculations. The area is different from the first two (Table 6 and 7) because the surface of the terrain that has been measured is concrete, while the first two are grasslands, hence the values are two and three times smaller.

Table 7. Compilation of calculations from the result file for area 2 in the local system.

Number of fixed points	**0**	**1**	**2**	**3**	**4**	**5**
Min. difference between series [m]	0.0107	0.0097	0.0094	0.0089	0.0088	0.0082
Max. difference between series [m]	0.0107	0.0107	0.0111	0.0109	0.0106	0.0099
Average difference between series [m]	0.0107	0.0104	0.0101	0.0099	0.0096	0.0092
RMSE [m]	±0.0155	±0.0153	±0.0151	±0.0149	±0.0146	±0.0144
Parameter ex	-0.000052	-0.000045	-0.00005	-0.00005	-0.000049	-0.000046
Parameter ey	-0.000061	-0.00006	-0.000061	-0.00006	-0.000063	-0.000066
Parameter H0	0.01786	0.01684	0.01673	0.01666	0.01646	0.01612

Table 8. Compilation of calculations from the result file for area 3 in the local system.

Number of fixed points	**0**	**1**	**2**	**3**	**4**	**5**
Min. difference between series [m]	0.0015	0.0013	0.0011	0.0008	0.0008	0.0009
Max. difference between series [m]	0.0015	0.0015	0.0016	0.0016	0.0016	0.0016
Average difference between series [m]	0.0015	0.0015	0.0014	0.0014	0.0013	0.0013
RMSE [m]	±0.00540	±0.00538	±0.00525	±0.00533	±0.00548	±0.00530
Parameter ex	-0.0001	-0.0001	-0.000098	-0.000097	-0.000096	-0.000095
Parameter ey	-0.00003	-0.00003	-0.000029	-0.000028	-0.000027	-0.000028
Parameter H0	0.00866	0.00865	0.00843	0.00831	0.00816	0.00811

Table 9. Compilation of calculations from the result file for area 4 in the local system.

Number of fixed points	0	1	2	3	4	5
Min. difference between series [m]	0.0029	0.0026	0.0025	0.0024	0.0022	0.002
Max. difference between series [m]	0.0029	0.0029	0.0031	0.003	0.0031	0.0028
Average difference between series [m]	0.0029	0.0028	0.0028	0.0026	0.0026	0.0025
RMSE [m]	±0.0057	±0.0056	±0.0055	±0.0054	±0.0053	±0.0052
Parameter ex	-0.000033	-0.000031	-0.000032	-0.000032	-0.00003	-0.000029
Parameter ey	-0.000054	-0.000052	-0.000052	-0.000051	-0.000046	-0.000045
Parameter H0	0.009155	0.0088	0.008737	0.008515	0.008088	0.007774

In Table 9 above, the results of calculations for the mixed area 4 - grass/asphalt are shown (the area with the smallest height difference). The calculations show that the average difference between the series is about ±27 mm, while the average error of the difference varies within ±5.5 mm. Both attributes show a decreasing trend with a growing number of fixed points. The ***ex*** parameter is ambiguous (it increases and decreases), the ***ey*** parameter increases and the ***H0*** parameter decreases in relation to the fixed points. The mean error is a value close to area no. 3, despite twice the average differences between the series of measurements.

Table 10. Compilation of calculations for four areas depending on the area slide.

	Area 1	Area 2	Area 3	Area 4
Terrain differences [m]	4.5	2.9	1	0.5
RMSE [m]	±0.0098	±0.0155	±0.0054	±0.0057
Parameter H0 [m]	-0.0149	0.0179	0.0087	0.0092

Taking into account the results of the calculations presented in Tables 6-9, it can be concluded (Table 10) that an attribute such as the error of the mean difference between the series of measurements is not accompanied by a decrease with decreasing height (terrain slide). The smallest turned out to be on the compact surfaces (area 3) and mixed ones (area 4). The largest error was calculated from data from area 2 (almost 3 times higher than in area 3 and 4). In area no. 1, this error is intermediate between extreme values and is about ±1 cm. The actual displacement, calculated on the basis of the analytical difference of the "0" function value and the values of the interpolated functions, vary from -14.9 mm (area 1) to +17.9 mm (area 2). In area 3 and 4, they are +9 mm. The ***H0*** parameter also does not accompany a decrease with the decrease in the value of the height difference. As in the case of RMSE, it turned out to be the smallest at small terrain differences, and the largest on large changes in the terrain elevation. This error points to the high accuracy of mapping the area of the terrain and fitting the DEM into the actual area.

5 Conclusion

The original program served to determine real vertical displacements between two measurements. The ***H0*** parameter perfectly shows these displacements between consecutive measurements and an artificially assumed theoretical "0" plane. It follows that theoretical values of plane parameters can be obtained as a result of calculations, with different combinations of non-fixed points (primary and secondary measurement) and theoretically fixed points. The ***H0*** parameter in areas 2, 3 and 4 displays a decreasing trend with the increase of the number of fixed points, while on area no. 1 (where the largest height difference and discrepancies between measurements were noted), it rises once and decreases once.

The calculated parameters show that the selection of an appropriate measuring technique for the measured terrain, and in principle its surface and height differences, is very important in order to obtain satisfactory results. As shown in this paper, the use of geometric leveling has proved to be sufficient to show that adding fixed points does not increase the accuracy of the obtained model, but it increases the accuracy of the surface parameters obtained. The use of the original calculation program has shown that it is possible to obtain pre-set parameters of the plane and comparing the obtained planes from these two measurements on survey points (not fixed in the field in reality).

In conclusion, the analysis of data in terms of displacements of non-fixed points permanently showed no significant displacements, discrepancies within their determined accuracy or slight changes in the shape of objects, mainly due to direct leveling as not very precise in relation to precise leveling. However, the use of this method seemed justified due to the high speed of data acquisition and processing, the choice of the measured objects, and also the purpose of obtaining this data (terrain surface).

This work was financed by the Polish Statutory Research Grant no. 0401/0123/17.

References

1. R. Baryła and J. Paziewski, Biuletyn WAT **2/2012**, 39-57 (2012)
2. A. Hermanowski, *Pomiary przemieszczeń pionowych* (Instytut Geodezji i Kartografii, Warszawa, 1993)
3. A. Kwinta, Zeszyty Naukowe Politechniki Śląskiej, Górnictwo, **278** (2007)
4. S. Oszczak, A. Wasilewski, Z. Rzepecka, A. Szostak-Chrzanowska, A. Chrzanowski, Proceedings 11th International Symposium on Deformation Measurements, Santorini, Greece, 25-28 May 2003 (2003)
5. A. Nowakowska and A. Rosa, *Pomiary przemieszczeń pionowych i poziomych budynku Starostwa Powiatowego w Opatowie* (GeoPixel, Wojskowa Akademia Techniczna, Warszawa, 2011)
6. H. Bryś and S. Przewłocki, *Geodezyjne metody pomiarów przemieszczeń budowli* (Wydawnictwo Naukowe PWN, Warszawa, 1998)
7. K. Czarnecki, *Geodezja współczesna* (Wyd. PWN, Warszawa, 2014)
8. A. Kwinta, A. Szeptalin, T. Gargula, [In:] Archiwum Geomatyki. Pomiary satelitarne w geodezji i budownictwie (Gdańsk, 2010)
9. T. Lazzarini, *Geodezyjne pomiary przemieszczeń budowli i ich otoczenia* (Państwowe Przedsiębiorstwo Wydawnictw Kartograficznych, Warszawa, 1977)
10. W. Prószyński and M. Kwaśniak, *Podstawy geodezyjnego wyznaczania przemieszczeń* (Oficyna Wydawnicza Politechniki Warszawskiej, Warszawa, 2006)
11. D. Gościewski, Materiały XX Jesiennej Szkoły Geodezji (Wrocław, 2007)
12. B. Hejmanowska, Archiwum Fotogrametrii, Kartografii i Teledetekcji, Vol. **17** (2007)
13. Cz. Suchocki, M. Damięcka-Suchocka, P. Błoch, M. Stec, Acta Sci. Pol., Geodesia et Descriptio Terrarum **12 (3)**, 17-26 (2013)
14. J. Wysocki, Przegląd Naukowy Inż. i Kształtowania Środowiska, nr **1 (31)**, 102–113 (2005)
15. C. Childs, *Interpolating surfaces in ArcGIS Spatial Analyst* (ESRI Education Services, 2012)
16. J.P. Mund, *Geospatial statistics and spatial data interpolation methods* (GIS'EM at Eberswalde, 2013)
17. S.M. Robeson, Cartog. Geog. Inf. Sys., **24 (1)**, 3-20 (1997)

Analysis of power lines span geometry based on TLS measurements

Andrzej Kwinta[1,*], *Karolina* Ważydrąg[1], and *Mariusz* Zygmunt[1]

[1]University of Agriculture in Krakow, Faculty of Environmental Engineering and Land Surveying, 21 A. Mickiewicz Av., 31-120 Krakow, Poland

Abstract. The inventory measurements of utility infrastructure are crucial issues in the field of engineering geodesy. Secure usage of power lines requires to monitor their geometry. According to classical approach to measurements of geometry of overhead power lines, it usually allows only to measure directly. Geodetic methods cannot be applied due to measuring limitations (lack of an appropriate area to conduct a measurement). Currently in engineering geodesy it is becoming more common to use methods of laser scanning as well as Unmanned Aerial Photogrammetric Survey. In the Paper, the procedure of determining of an overhead power line's geometry is presented on the example of an one span measured with Terrestrial Laser Scanning methods. Field measurements were conducted by using laser scanner. Afterwards the point cloud was processed in order to receive the geometry of conductor. Finally, results were presented is tabular and graphic form.

1 Introduction

The existence of societies in the strongly urbanized and industrialized areas is based on the usage of technical infrastructure. In modern times it is impossible for the economy as well as the society to operate without various media delivered with the (dependable) technical infrastructure [1]. It can be told that the area is entangled with overhead, terrestrial and underground networks of utility infrastructure. The proper operation of those infrastructure is significant for well being of societies.

Constructions of technical infrastructure are crucial elements of human living space [2]. Those constructions determine the comfort and the security of men existence. People get used to benefits that are provided by media delivered with utility infrastructure. A present-day city inhabitant do not imagine the existence without access to various media. Therefore, it is important to control the condition of utility infrastructure form the operational as well as the geometrical point of view.

Power lines transmit electrical energy from generating sites to power consumers [3]. Usually (due to harmfulness of radiation) overhead power lines are being constructed in undeveloped areas, very often such as woodlands. In those situations capabilities of conducting measurements by using classical geodetic methods are limited and the only applicable way of measurement is a direct one or a one with usage of remote sensing methods. Lately ALS methods and low-altitude UAV photogrammetry are becoming more commonly used to conduct power lines' measurements.

The implementation of GNSS measuring systems as well as the development of laser scanning technologies undoubtedly facilitate the conducting of geodetic measurements. In the Paper is presented the issues of determining a power line's span geometry based on laser scanning measurement. Therefore, the method of performing TLS measurements and calculating its results is not the subject of this work. Measurement TLS has been described only to a minimal extent. Presented issues are illustrated with the example. The example uses measured and processed data from TLS measurements in the form of a point cloud (point coordinates).

2 The geometry of overhead power line

An overhead power line consists of many elements, which undergo measurements and analyses [3]. The power line is an element of the infrastructure, which transmits electrical energy from a generating site to end consumers (users). A single power line can be divided into spans, which are sections of cable between two hook points. According to parameters of a power line, it can be constructed differently and a line's geometry can differ significantly.

Single conductor between two support points can be considered as freely hanging rope, which the weight has an effect on. In such an instance the cable takes shape of a catenary. The freely hanging cable can be also described with a quadric polynomial. Below in the picture 1, basic geometric parameters of single cable are presented.

* Corresponding author: andrzej.kwinta@ur.krakow.pl

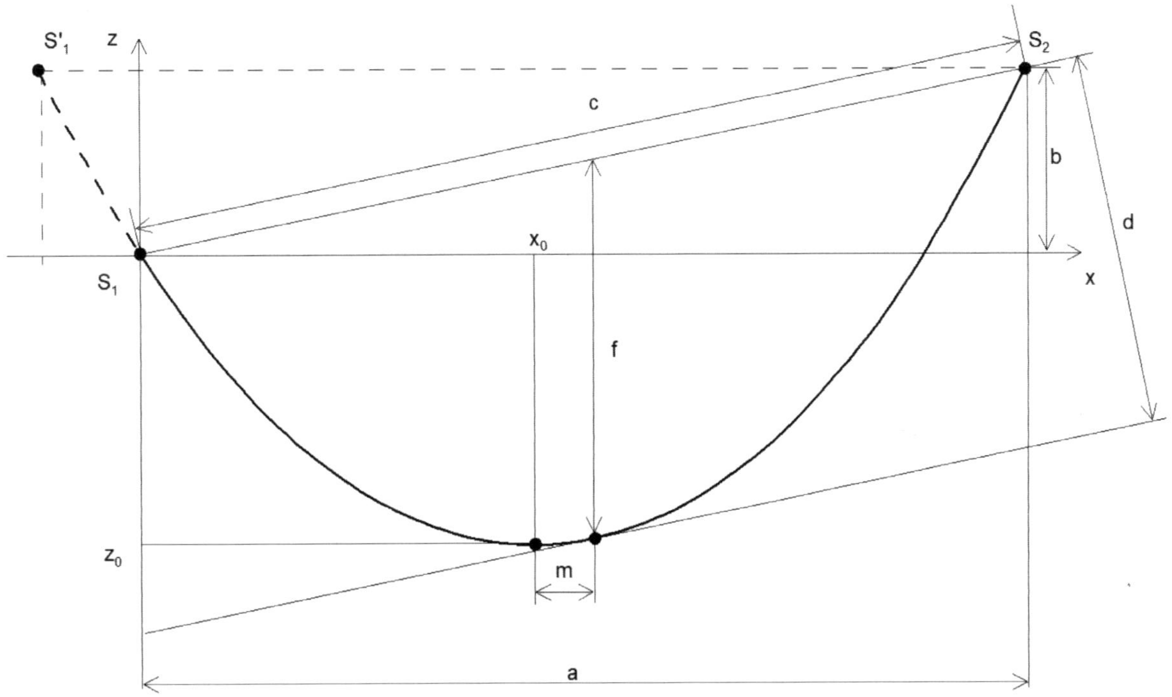

Fig. 1. Geometric parameters of span of freely hanging cable.

According to the Fig. 1 following geometric parameters can be distinguished:

- span length (*a*) – a horizontal distance between axes of adjacent supporting structures (cable support points)
- chord (*c*) – a straight line segment, whose endpoints are cable support points
- sag (*f*) – the biggest vertical distance between a cable and a chord
- slope (*b*) – a vertical distance between two support points of one cable of an inclined span
- eccentricity (*m*) – of an inclined span; a distance between the lowest point of a cable and a centre of a span
- devilation (*d*) – the biggest distance between a curve of hanging cable and a chord; measured perpendicularly to a chord
- length (*a*) – a length of a cable between two insulators
- the lowest point (X_0, Z_0) – coordinates of the lowest point of a cable

There are two theoretical equations which describes a cable geometry [4, 5]:

- catenary

$$z(x) = k \cdot \cosh\left(\frac{x - x_0}{k}\right) + S \tag{1}$$

where:
k is a catenary shape parameter
x_0 – a coordinate of the lowest point
S – a translation in vertical plane

- parabola

$$z(x) = px^2 + qx + r \tag{2}$$

where *p*, *q*, *r* are parameters of parabola (*q=0* when a parabola is a polynomial approximation of a catenary).

On the basis of abovementioned formulas and the Fig. 1, following equations for geometric parameters of single cable can be formulated according to adopted theoretical function:

- span length

$$a = x_{max} - x_{min} \tag{3}$$

- slope

$$b = \left| z(x_{max}) - z(x_{min}) \right| \tag{4}$$

- chord

$$c = \sqrt{a^2 + b^2} \tag{5}$$

- sag
  - catenary

$$f = k\left[\frac{bg}{a} - \cosh(g)\right] + \frac{bx_0}{a} - S \tag{6a}$$

where $g = arcsinh(b/a)$

  - parabola

$$f = \frac{(b - aq)^2}{4a^2 p} - r \tag{6b}$$

- eccentricity
  - catenary

$$m = k \cdot \operatorname{arcsinh}\left(\frac{b}{a}\right) \tag{7a}$$

  - parabola

$$m = \frac{b}{2ap} \tag{7b}$$

- devilation
  - catenary

$$d = \frac{a \cdot f}{c} \tag{8a}$$

  - parabola

$$d = \frac{(b-aq)^2 - 4a^2 pr}{4ap\sqrt{a^2+b^2}} \tag{8b}$$

- length
  - catenary

$$\alpha = k\left[\sinh\left(\frac{a-x_0}{k}\right) - \sinh\left(\frac{-x_0}{k}\right)\right] \tag{9a}$$

  - parabola

$$\alpha = \frac{t\sqrt{t^2+1} + \operatorname{arcsinh}(t) - q\sqrt{q^2+1} - \operatorname{arcsinh}(q)}{4p} \tag{9b}$$

where $t=2ap+q$

- the lowest point
  - catenary

$$X_0 = x_0 \tag{10a}$$

$$Z_0 = k + S \tag{11a}$$

  - parabola

$$X_0 = -\frac{q}{2p} \tag{10b}$$

$$Z_0 = \frac{4pr - q^2}{4p} \tag{11b}$$

As it can be seen both equations (1) and (2) have three parameters, that is why to unambiguously determine equations, three points of measurement of a cable are needed. Taking into consideration span's geometric parameters, two opposite support points of a conductor and one point on a conductor has to be measured.

3 Methods of measurement

Measurements of overhead power lines are, as a rule, conducted with geodetic methods. Not long time ago basic and most popular method was measuring of vertical and horizontal angles as well as distances. Currently methods based on low-altitude photogrammetry and mobile scanning systems (aerial as well as automotive). In some cases (a measurement of deflection only) other non-geodetic (direct) measurement methods, which requires climbing up on supporting structures, are being used.

Regarding direct measuring methods, following are being used [6, 7]:

- driven oscillations method (a cable oscillation is driven with shudder and a time is measured),
- surveying targets method (a measurement of an vertical angle to cable's tangent),
- levelling rods method (outlining a tangent in the point of maximal sag).

Following geodetic methods are being used [8-12]:

- angular intersection (conducted in vertical plane),
- trigonometric method (a measurement of vertical and horizontal angles as well as distances),
- aerial photogrammetric survey (classical and low-altitude UAV),
- laser scanning (aerial, mobile, terrestrial),
- remote sensing methods,
- GPS measurement.

Each of methods enlisted above has its advantages and disadvantages, what determine the choice of most suitable measurement method by taking into consideration actual needs, measurement's conditions as well as a geometry of measured object.

Laser scanning technology, which a few years ago was considered to be the most innovative and modern one, currently has become the primary measurement method used in geodesy. A common access to measuring instruments as well as computer software has led to ubiquitous application of this method to measuring any of engineering objects. One of basic parameters connected with conducting a measurement with laser scanning method is choice of appropriate measurement points' density in accordance with an object's geometry as well as requirements to a measurement accuracy. Too low density of points causes a decline in accuracy of determining a geometry of object, but in the same time this leads to speeding up measurements and calculations. On the other hand, too high density of points causes significant extension of measurement time and leads to difficulties in reprocessing obtained data. A reprocessing of measurement outcomes requires performing some operations, which are dependent on an available computer software.

4 Data evaluation

As a result of the laser scanning method measurements, one or more point clouds are obtained. Those point clouds have to be reprocessed in order to calculate geometrical elements of a measured object. Certain operations have to be performed to extract measured points of measured object from a point cloud and afterwards calculate object's geometrical elements [13-19].

Producers of measuring instruments and computer software are delivering solutions for reprocessing data in automatic or semi-automatic form [20], but for this Paper the manual reprocessing of outcomes has been conducted.

In order to obtain geometric data on the basis of the laser scanning method measurement, the following calculation steps have to be taken:

1. Downloading measurement data from measuring instrument to computer
2. Merging of point clouds (if there are more than one) and registration of data in a presupposed spherical coordinate system
3. Separating the object from point cloud, extraction of single cables in the analysed span of power line
4. Transforming the measuring three-dimensional coordinate system into two-dimensional coordinate system connected with the certain cable in the certain span of the power line (the cable's hook point at the origin of coordinate system, the axis of coordinate system in the same vertical plane with the cable's axis of geometry)
5. Estimated determining of the cable's theoretical equation (e.g. a catenary or a parabola)
6. Direct determining of parameters of the cable's theoretical equation in accordance with the Least Squares method (the least-squares function approximation)
7. Calculating conductor's parameters (*vide* Part 2 of the Paper)
8. Analysing the accuracy (determining of errors of calculated equations' parameters and geometric parameters)
9. Transforming results into spherical coordinates system
10. Presenting results of measurements and calculations (in tabular and graphic forms)

Abovementioned steps connected with calculations are automated to some bigger or smaller extent depending on a computer software used for reprocessing, but the verification of calculations' correctness is necessary by the operator. Authors of the Paper want to clearly state that the automation of calculations is desired, but on the other hand it is essential to verify outcomes of particular calculation stages. This publication is mainly about steps 5-10 of the presented procedure. The authors assumed that steps 1-4 of the procedure cover issues that will be the subject of separate publication.

Taking into account the density of scanning, and therefore the volume of data (size of a point cloud), conducting calculations might require high-speed computers and hard discs of high capacity and moreover calculations might be highly time-consuming. In the next section of the Paper, the process of conducting calculations of the measured power line's span is described.

5 Example

The measurement presented in the Paper were conducted for scientific purposes, that is why the high density of measurement points has been obtained [21]. This set of data has been collected for the needs of various analyses. In the Paper only the part of data has been used.

The measurement has been conducted with the scanner Leica Scan Station P30 [22] in Krakow-Mydlniki (Fig. 2). The measured object was a span of 220 kV power line of 300 meters range. The power line consists of six working conductors and two ground wires. In the Fig. 2, the fragment of measured power line is presented.

Fig. 2. The fragment of the measured power line and the set measuring instrument.

Measurements were conducted on three positions with the use of six adjustment points. Adjustment points were measured classically and their coordinates were marked out in the local coordinates system. The single working cable (marked with red in the Fig. 3) was chosen for analysis, which the point cloud of 903 516 points was selected for, what gives the average density of 3 points per millimetre of the cable.

Fig. 3. Visualisation of the point cloud with the analysed cable marked in red colour.

The data from the measuring instrument was downloaded and merged with LEICA CYCLONE software [23]. Particular point clouds from different measurement positions were combined into one set and saved as '*.pod' text file.

In order to analyse outcomes, the separation of data was needed and the point cloud had to be divided to extract particular cables, what allowed to determine their geometric parameters. This reprocessing was conducted with Bentley DESCARTES software [24]. In the Fig. 3 the single cable of the analysed power line's span picked for the further analysis is marked (in red colour). Only one single cable was chosen due to the fact that the analysis of all cables did not give any additional information in this particular case. It is crucial during choosing data to correctly identify hook points of the cable on the supporting structure, because this in fact determines span's geometric parameters.

To conduct the calculation the original software has been created in DELPHI programming language in RAD Studio application [25]. Data transformations, estimated calculations, precise calculations as well as determination of approximation's errors were done with this software. Firstly, as results of conducted calculations theoretical equations of curves approximating the electric cable were determined by calculating estimated values and later precise ones of equations' parameters:

– the function of a catenary

$$z = 1136.296\cosh\left(\frac{x-136.290}{1136.296}\right) - 1144.573 \qquad (12)$$

– the function of a parabola

$$z = 0.0004414x^2 - 0.120381x - 0.076 \qquad (13)$$

On the basis of determined theoretical equations and formulas (3)-(11) presented in part 2 of the Paper the span's parameters have been calculated and outcomes are compiled below in the Table 1.

While analysing outcomes compiled in the Table 1 it had to be stated that values of geometric parameters of cable's span received with the approximation theory were very similar for either the catenary as well as the parabola. The maximum difference equalled around 0.1 m and concerned the eccentricity, what was caused by the distinct location of the lowest point (difference of 0.073 m) according to both of theories. Regarding the sag, the difference was of 0.002 m, what was the negligible value, while taking into account the total value of the sag.

Along with the approximation of outcomes of span's measurements, not only theoretical functions (12) and (13) were calculated, but average errors of approximation were determined. For the function of a catenary the average error of approximation equalled 0.035 m, while for the function of a parabola the average error equalled 0.039 m. When comparing received outcomes it had to stated that both values were similar and both theoretical functions described the measured power line's cable correctly.

Numerical results of calculations were saved as appropriate text files, which were used to create the visualisation in a graphic program. The Bentley MICROSTATION software [26] was used to create the data visualisation. The point cloud obtained from measurements uploaded to the program as the '*.pod' text file (default layer of picture) and later appropriate calculations' results were placed on the next layers of the picture. During calculations the proper textual structure of files was conferred and therefore calculations' results in loadable form were possible to have been included on layers of the picture.

For each of theoretical curves (the catenary, the parabola) following layers of the picture were created:

– the layer of the theoretical curve – contain the equation in textual form and the theoretical curve marked with appropriate colour;
– the layer of cable's parameters – containing the numerical sheet of geometric parameters of the cable's span and graphic form of appropriate points as well as line segments in the picture;
– the layer of the errors' analysis – containing textual description of cable's approximation errors for particular points marked in different colours. The division of errors into 10 categories (every 7 mm) was adopted and to each category different colour was assigned (Fig. 4).

As the example of results presented in graphic form, in the picture below the snapshot of the default layer and the layer of the errors' analysis for the cable selected for calculations. Also, the small part of the analysed cable is magnified (Fig. 4).

Table 1. Geometric parameters of cable's span (all results are presented in meters).

Parameter	Catenary	Parabola	Difference
Span length	298.019	298.019	0.000
Slope	3.346	3.327	0.018
Chord	298.037	298.037	0.000
Sag	9.879	9.877	0.002
Eccentricity	12.756	12.647	0.109
Devilation	9.878	9.876	0.002
Length	298.892	298.894	-0.002
The lowest point (X_0)	136.290	136.363	-0.073
The lowest point (Z_0)	-8.270	-8.284	0.007

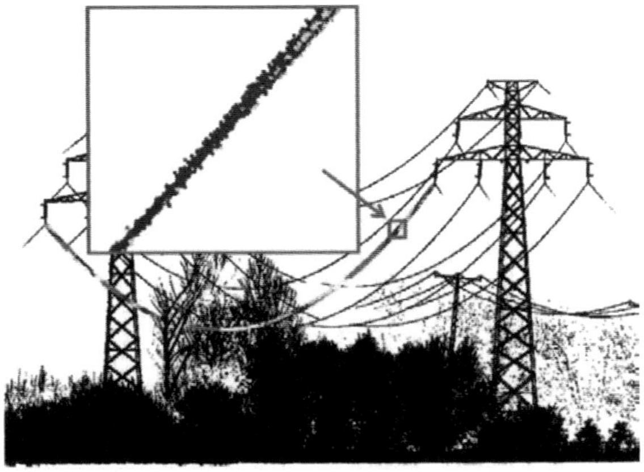

Fig. 4. Visualisation of the point cloud with the analysed cable marked in red colour.

Such a big accumulation of data as used in example presented required the usage of the appropriate computer, due to the fact that either calculations as well as the visualisation required the large computer memory and the good graphics card, but even then it took a long time to receive the results of calculations and to refresh the picture.

Presented in the Paper the example of the analysis was initial and preparatory, moreover it is only one of stages for further processing. Further actions would lead to automation of the calculations' process and the visualisation of results. Algorithms obtained would reprocessed into VisualBasic language (MicroStation Basic) [27], what would allow to use them directly from the MICROSTATION level.

6 Summary

The usage of the laser scanning technology in geodesy is not a novelty anymore and become the common measuring technology. Decreasing prices of measuring instruments and growing access to the professional software leads to increasing interest in this measuring technology.

Determining of the overhead power liens' geometry is the issue quite commonly encountered in the geodetic practice. The required accuracy as well as the scope of measurements need to be defined dependent on the aim of measurements and analyses. In some cases, it is not possible to conduct classic measurements due to the terrestrial limitations (lack of appropriate visibility), in the same time usage of the Aerial Photogrammetric Survey or the mobile laser scanning are not economically justifiable. In those cases, the best possible solution is to apply the terrestrial laser scanning method.

It should be emphasized that the methodology of performing TLS measurements and calculating its results was not the subject of this work. Therefore, the results of TLS measurements have been described to a minimum without details. The description of TLS measurements will form the basis of another article

In the Paper, succeeding steps of calculations connected with the analysis of measurements' outcomes and the visualisation of them have been presented. The mathematical relations used to describe the geometry of a hanging cable have been depicted depending on the adopted approximate theoretical function. Calculations and the presentation of results have been illustrated with the example.

Results of approximation for two theoretical curves, those are a catenary and a parabola, have been compared. Obtained results have confirmed that both of those theoretical functions can be applied to calculations.

In the Paper has been used the original software (special made) for the analysis and the visualisation of a overhead power lines geometry. Subsequent steps of the analysis can be automated but the particular caution has to be kept and results obtained have to be verified. The quality and the accuracy of obtained outcomes extensively depends on correctly planned and conducted measurements as well as on correct calculations.

References

1. M. Antrop, Landscape and Urban Planning, **67**, Issues 1–4, 9-26 (2004)
2. M. Hopkowicz, S.M. Rybicki, Geomatics, Landmanagement and Landscape, **1**, 19-33 (2014)
3. F. Kiessling, P. Nefzger, J.F. Nolasco, U. Kaintzyk, *Overhead Power Lines: Planning, Design, Construction* (Springer, 2014)
4. A. Hatibovic, Geomatics, Electrotehnica, Electronica, Automatica, **60**, 3, 9-14 (2012)
5. Task Force B2.12.3, *Sag-Tension Calculation Methods for Overhead Lines* (CIGRE 324 Technical Brochures, 2016)
6. J. Migas, J. Hauschild, *Prace geodezyjne przy budowie linii i stacji elektroenergetycznych* (ELBUD, Kraków, 1996)
7. D.A. Douglass, R. Thrash, *Sag and Tension of Conductor* (Taylor and Fransis Group LLC., 2006)
8. A. Wehr, U. Lohr, ISPRS Journal of Photogrammetry and Remote Sensing, **54**, 68-82 (1999)
9. P. Amon, U. Riegl, P. Rieger, M. Pfennigbauer, *UAV-based laser scanning to meet special challenges in lidar surveying* (Geomatics Indaba Proceedings, 2015)
10. L. Matikainen, M. Lehtomäki, E. Ahokas, J. Hyyppä, M. Karjalainen, A. Jaakkola, A. Kukko, T. Heinonen, ISPRS Journal of Photogrammetry and Remote Sensing, **119**, 10-31 (2016)
11. S. Kamboj, R. Dahiya, International Journal on Electrical Engineering and Informatics, **3**, 3, 268-277 (2011)
12. P. Ćwiąkała, Prace Naukowe Instytutu Górnictwa Politechniki Wrocławskiej. Konferencje, **120**, 49, 89-97 (2007)
13. J. Kremer, *Power Line Mapping: Data Acquisition with A Specialized Multi-Sensor Platform*

(Photogrammetric Week '11, Wichmann/VDE Verlag, 2011)

14. T. Melzer, C. Briese, *Extraction and modeling of power lines from ALS point clouds* (Proceeding of 28th OAGM Workshop, Austrian Association for Pattern Recognition, Osterreichische Computer Gesellschaft, Hagenberg, Austia, 2004)
15. B. Kwoczyńska, J. Dobek, Journal of Ecological Engineering, **17**, 4, 65-74 (2007)
16. P. Axelsson, ISPRS Journal of Photogrammetry and Remote Sensing, **54**, 4, 138-147 (1999)
17. B. Song, X. Li, Neurocomputing, **129**, 350-361 (2014)
18. G. Yan, C. Li, G. Zhou, W. Zhang, X. Li, IEEE Geoscience and Remote Sensing Letters, **4**, 3, 387-391 (2007)
19. Y. Jwa, G. Sohn, H. B. Kim, The International Archives of the Photogrammetry, Remote Sensing and Spatial Information Sciences, **XXXVIII**, Part 3/W8, 105-110 (2009)

Modelling of underground objects and geological-tectonic structures from a point cloud: application of the developed models in studies of geodynamic processes

Damian Kasza[1,*]

[1]Wrocław University of Science and Technology, Faculty of Geoengineering, Mining and Geology, 27 Wyb. Wyspiańskiego St., 50-370 Wrocław, Poland

Abstract. At present, techniques of terrestrial laser scanning are one of the most popular survey methods. They allow for fast and non-invasive data acquisition even up to several tens of thousands points per minute. Apart from high precision and measurement efficiency, one of its assets is the possibility of working in objects with natural or artificial light. Due to this, the method may be successfully used in activities related e.g. with the surveying of the present state of underground excavations. The basic product derived from laser scanning measurements is a point cloud. This set is the input data for the construction of a 3D object with a closed surface, e.g. a mesh network. This paper presents the results of inventory works performed in a complex of underground excavations below Książ Castle in Lower Silesia (SW Poland). Based on a point cloud acquired during the measurements, a 3D model of the excavations was constructed. Open source software – MeshLab – was used in the modelling. The high-resolution, calibrated metric model of the underground excavations allowed for a macroscopic recognition of the position of the main tectonic structures cutting the space of the underground excavations. The obtained model was supplemented with the results of earlier geodetic and geological surveys, performed in 2013-2015. The generated model of the geological structure in the study area is an element supporting the interpretation of kinematic activity of the rock massif registered by the instruments of the SRC PAS Geodynamic Laboratory in Książ.

1 Introduction

Studies of the natural environment are a key to recognise the mechanisms of processes taking place within it. This particularly applies to Earth Sciences: geology, geophysics, geochemistry, etc. Continuous measurements of displacements performed by permanent GNSS receivers or registered by seismological observatories are good examples. The data that they supply are the basis for the analysis and interpretation of phenomena such as displacements of the upper layers of the lithosphere or effects caused by earthquakes.

However, individual study results are never sufficient enough. Presentation of the modelled phenomena requires a virtual image of the environment in which the given processes take place. This requires the use of digital models of land surface or the upper parts of the lithosphere from soundings or from analysis of material from deep boreholes.

In time of developing computer tools, projections of numerical surfaces of Earth's globe constructed with satellite photographs, data from laser scanning and photogrammetric images using unmanned aerial vehicles attain an increasing number of applications. Terrestrial laser scanning belongs to such new techniques. Being used in an increasingly wider spectrum of possibilities [1, 2], it allows for obtaining extremely precise reflections of the scanned objects/surfaces in a virtual space.

The paper presents the results of studies performed in the underground object "Książ" (Lower Silesia, SW Poland). They were focused on constructing a model of the underground excavations with determining the position of tectonic structures, i.e. discontinuity surfaces visible in their space.

2 Study area

The complex of underground corridors "Książ" was constructed directly below the courtyard of the Książ Castle near Wałbrzych (Fig. 1). The corridors were excavated during the Second World War in 1943-1945. According to the existing documents, the object was part of a Nazi secret military project with the cryptonym S3 "Riese" [3-5], which was aimed at the construction of a series of underground objects in Książ and in the Sowie Range (the Sudetes). Their destination remains unknown, most probably they were supposed to function as laboratories and/or armament objects.

* Corresponding author: damian.kasza@pwr.edu.pl

https://doi.org/10.1051/e3sconf/20185500014

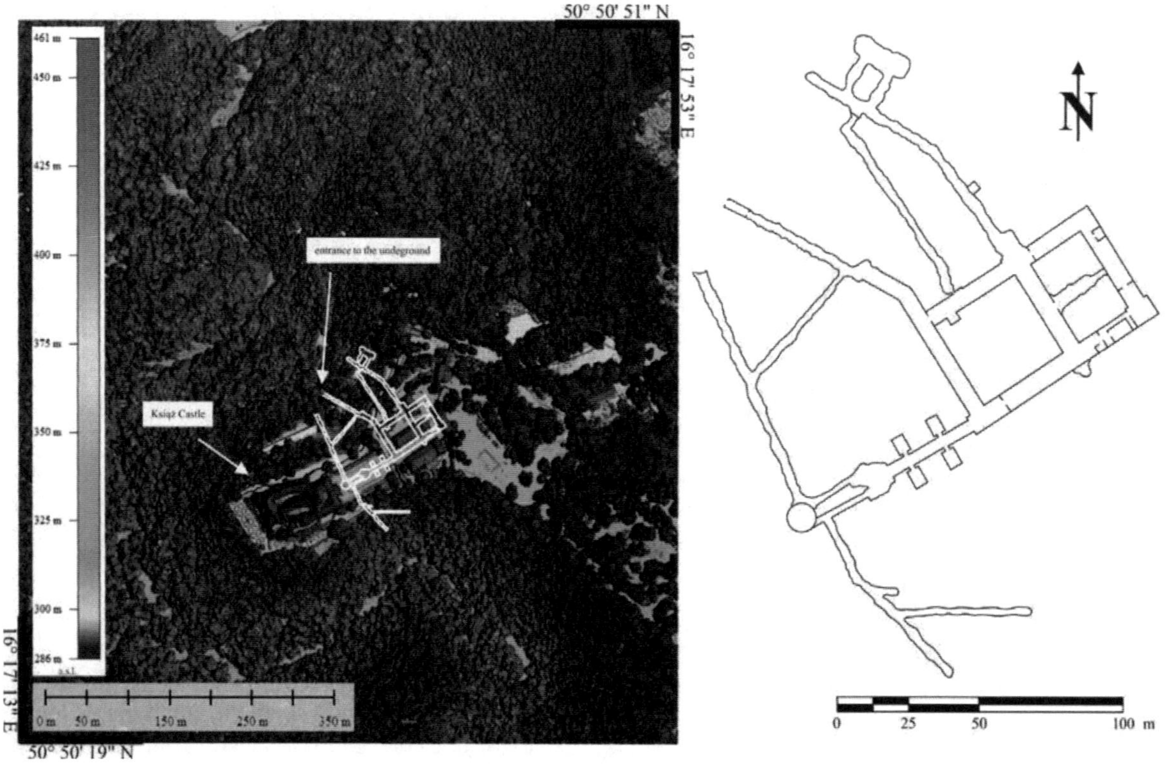

Fig. 1. Plan of underground corridors of the "Książ" object [6].

Since the 1970-ties, the underground corridors at Książ play the role of a laboratory belonging to the Polish Academy of Sciences (PAS). At present it houses the Seismological Observatory of the Institute of Geophysics (Polish Academy of Sciences) and the Geodynamic Laboratory, belonging to the Space Research Centre of the Polish Academy of Sciences (GL SRC).

The "Książ" object is located in the centre of a structural unit known as the Świebodzice Basin. The entire space of the corridors was excavated within a sedimentary rocks known as the Książ Formation, which is Late Devonian to Early Carboniferous in age. The Formation includes conglomerates, with the clasts dominated by gneisses, granite-gneisses and light grey granites [7-12].

The unique properties of the Książ massif (non-permeable rocks, dense fault network) allow for registering subtle effects of tectonic phenomena by sets of GL SRC measurement equipment: permanent GNSS stations, quartz horizontal pendulums, water-tube tiltmeters, vacuum interference extensometer and TM-71 crack gauge [11, 13-21].

3 Measurements with terrestrial laser scanning technique

The concept of measurements using terrestrial laser scanners is based on the measurement of distances and angles (both horizontal and vertical) in relation to the measurement point. Based on these three elements the position of the measured point in space is determined in the scanner local system. By tying the scanner measurement site to the grid point, it is possible to align the obtained measurements to global coordinates. A point cloud with a given density is obtained from the measurements. Single measurement sites of the scanner may be connected by equivalent reference points (at least 3). The reflection coefficient is an additional data obtained during measurement with a laser scanner; its value depends on the type of material (including physical properties), of which the measured object (point) is composed.

A Leica ScanStation C10 impulse scanner was used during the survey in Książ (Tab. 1). Earlier stabilised points of the horizontal grid were used as reference points for connecting the measurement sites (Fig. 2) [19, 22]. Due to the large dimensions of the surveyed excavations and an adequate distribution of the grid points, the average errors of the position of the measured points were minimized. Connection of particular sites was made in the Leica Cyclone environment. The resulting point cloud was aligned to the PL2000 zone 5 system through four points located in front of the entrance to the underground excavations (coordinates determined based on RTK GPS measurements).

4 Data analysis

A large amount of the collected data and high expectations regarding the specificity of the constructed model determined the need to seek a highly effective method of data processing allowing for precise reconstruction of the scanned surfaces. Open source software – MeshLab [23] – was selected for work with the point cloud. Its assets included e.g. a large number of data formats – both input and output data, several algorithms of constructing 3D models, processing millions of point datasets and a wide array of editing tools. Among the methods of constructing models available in the software functions, the *out-of-core Poisson Surface Reconstruction* was used in the analysis.

Table 1. Scanning parameters.

Number of sites	43
Scanning resolution	5 x 5 cm on a distance of 50 m
Scanning duration in each site	~ 7 minutes
Volume of point cloud	149 371 264 points

Fig. 2. Scanner measurement site.

4.1 Poisson Surface Reconstruction Algorithm

Solution of the problem of mesh type surfaces from a point cloud has been described in detail by Kazhdan et al. [24] and Bolitho et al. [25]. These authors presented the reconstruction of a surface from a point cloud as a solution of the Poisson problem in space. It is based on the transformation of a point set into a spatial continuous vector field, followed by finding a scalar function, whose gradients best fit to the vector field, and constructing a specific isosurface. The advantage of this method is data processing as a single dataset (without subdivision into local zones and their further joining to obtain a complete model). Besides, the algorithm is resistant to errors derived from the imprecision of stacking the scanner sites or lack of data (points) in some regions [26].

During the surface reconstruction process, based on the Poisson algorithm, MeshLab presents the data structure as an octree with a given depth. The larger depth value results in a more precise output model. This results with increasing the duration of the process and utilization of the computer resources. Application of multi-level data streaming during surface construction using the *Poisson Surface Reconstruction* algorithm, presented by Bolitho et al. [27], has allowed for work with large datasets but has generally increased the effectiveness of the process at restricted resources (processing power, computer memory) without loss of precision of the constructed models.

4.2 Construction of a 3D model for the "Książ" object

A ready-made packet of raw measurement data required indispensable initial processing before the construction of the mesh model. Points located beyond the underground excavations (fragments of land surface, vegetation, etc.) and noise, i.e. points located in direct proximity of the scanned excavations but not being part of them, were removed manually in the first step. Additionally, *the Poisson-disk sampling* [28] filter was applied to the dataset, which allowed for unifying the point cloud (the minimal distance between the neighbouring points was determined at the level of 1 cm). The result of this operation was decreasing the point density in direct proximity of the measurement sites, and therefore increase of the degree of sustainable distribution of points of the scanned surface.

After these operations, precision of the constructed mesh models was tested with regard to the applied parameters of the Poisson Surface Reconstruction algorithm. The most important among them, Octree depth, determined the degree of recurrent subdivision of the octree structure, and therefore larger specificity of the constructed mesh model. For values of this parameter exceeding 14, significant improvement of the quality of the obtained models was not observed (Tab. 2). The remaining coefficients influencing the degree of computer resources utilization, the minimal number of points visible in the tree branch and the method of isosurface compensation have been retained with default values (8; 1; 1).

The process of mesh surface construction for depth coefficient at 14 lasted for about 2 h (the computer used was a 6th generation Intel® Core™ i7-6700HQ processor, 32 GB RAM 1600 MHz, SSD hard disk). The constructed mesh surface was composed of almost 95 million triangles. The effect of this operation is presented in Fig. 3.

 https://doi.org/10.1051/e3sconf/20185500014

Table 2. Quantitative results of mesh model reconstruction depending on the applied depth value parameter.

Number of points	Depth	Number of triangles	Duration of model construction [hours:minutes:seconds]	Utilisation of memory resources (peak memory) [MB]
61 698 648	10	407 910	00:08:25	271.54
	11	1 706 185	00:09:10	279.17
	12	7 024 488	00:14:13	295.54
	13	27 986 164	00:43:39	855.00
	14	94 932 204	01:56:00	2,740.00
	15	94 598 409	02:50:49	1,950.00
	16	94 403 184	03:31:50	3,510.00
	17	94 316 872	06:19:50	5,870.00
	18	94 235 574	10:34:12	7,620.00

Fig. 3. Visible fragment of the model of entrance adit with visible surfaces of concrete casing (grey) and unprotected part of corridor (black).

5 Results

The mesh structure constructed from the measurement points is a metric product approximately corresponding to the present state of the excavations. Based on the assumption of a precise and correct reflection of the structure of the corridors that have not been rebuilt, an attempt was made to recognise discontinuity zones, whose presence was determined during earlier structural investigations and observation of sidewalls during scanning of the excavations.

Correlation of the constructed model with the results of *in situ* measurements of the tectonic structures [12, 22] has resulted in distinguishing the position of several tens of discontinuity zones. Application of Autodesk 3DS Max software for each dislocation zone resulted in determining study sites in the 3D space. The closed sites created Bézier type surfaces (Fig. 4) [22, 29], corresponding to fault surfaces cutting the excavation space.

For a fast acquisition of data on the azimuth of fault orientation and dip angle of a given fault, a plane had to be aligned to the given surface. Solution of this issue required the use of an algorithm worked out by Kasza et al. [19]. It assumes calculating a normal plane (1) in a Cartesian system, where the sum of squared distances between points (representing the fault position) to this plane is the smallest (2) [19]:

$$0 = Xcos(\alpha) + Ycos(\beta) + Zcos(\gamma) - p \qquad (1)$$

where p is the distance of the plane to the beginning of the measurement system;

$$S = \sum_{i=1}^{N} \delta_i^{\,2}(x_i, y_i, z_i) \qquad (2)$$

where N is the number of measurement pickets for a given fault (peaks of a surface drawn in the 3D space).

The peaks of a 3D surface drawn for a given fault were attained as measurement points, whereas the beginning of the reference system was one of the three GPS points used for the calibration of the scanned object. As a result, values of dip and strike of planes representing the faults were calculated for each of the determined surfaces (Fig. 5). Calibration of the obtained image of fault orientation in a 2D space with earlier reports [19, 22] allowed to confirm the directional compatibility of the obtained projections for a given population of faults in the Książ underground space.

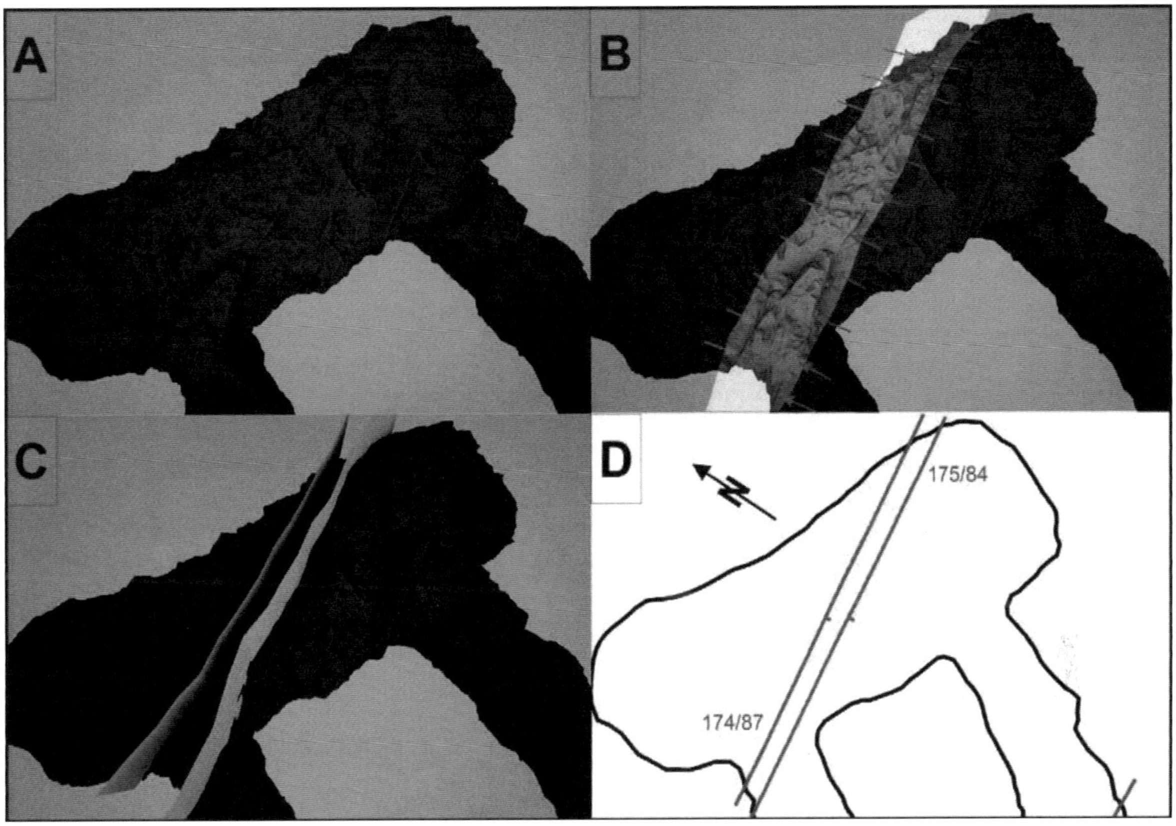

Fig. 4. The so called "Guardroom": A) 3D model as a mesh type structure; B), position of identified discontinuity surfaces; C) generated models of the discontinuity surfaces (higher range surfaces); F) fragment of constructed geological sketch with marked discontinuity zones and calculated values of directional parameters.

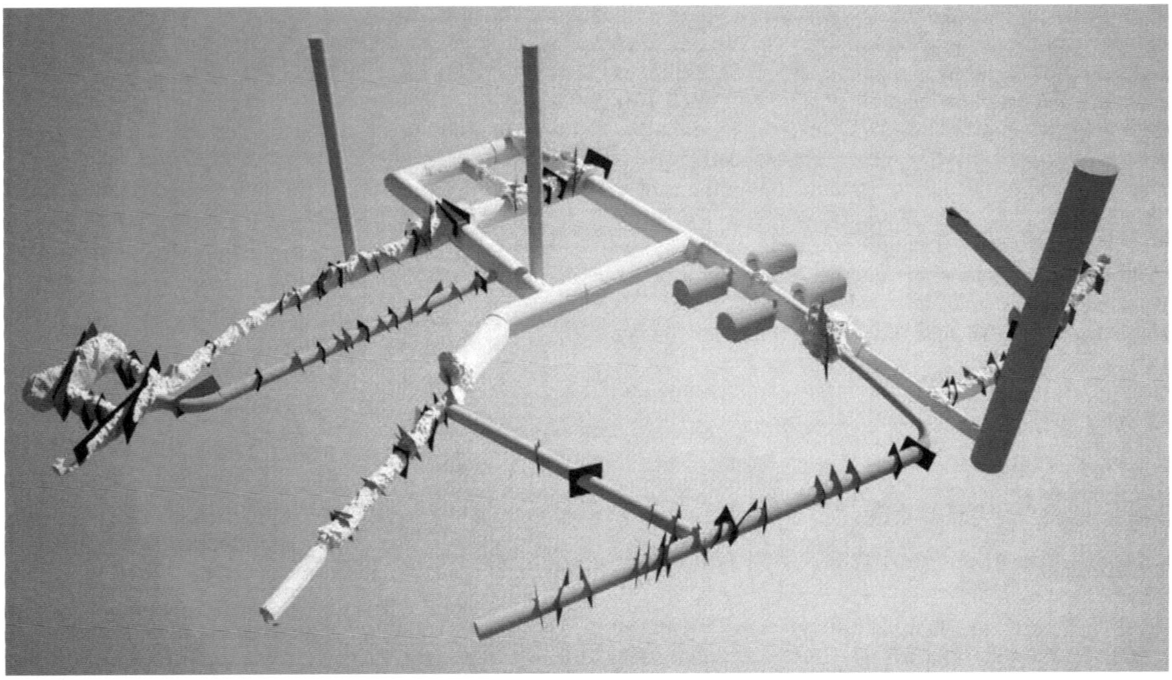

Fig. 5. 3D model of underground excavations below Książ Castle (grey and yellow) with marked position of the main discontinuity zones (blue).

6 Discussion

Use of input data in form of a point cloud and MeshLab software has allowed for constructing a high-resolution metric model of the underground excavations at Książ. The applied algorithm used in the reconstruction of a mesh surface – *Poisson Surface Reconstruction* – turned out to be optimal at application of reconstruction depth at 14 (Tab. 2).

The further part of the study – generation of the course of surfaces marking the discontinuity zones – has allowed to reconstruct the position of the main tectonic zones cutting the space of the underground excavations. Comparison of these results with earlier tachymetric surveying [22] has confirmed their compatibility.

The constructed model is a base for derived products such as maps or profiles. The large amount of data types, which may be recorded from the level of the MeshLab and 3DS Max environments allows for their use and further processing using other software, both commercial and open source.

7 Conclusions

Terrestrial laser scanning presently belongs to one of the fastest and most effective methods of acquiring information on the topography of underground objects. It allows for a precise method of reconstructing the scanned objects in a virtual space, and, moreover, their reference with existing numerical reports.

The presented method of data processing and analysis indicates its usability in many aspects. It is a significant support in surveys related with geological documentation. Due to a unique possibility of observing the massif from inside (penetration of excavations) and conducting the presented surveys, it was possible to construct a digital spatial geological-tectonic model of the study area. Such projection of the massif in the region of Książ may become a valuable database for geodynamic investigations in GL SRC PAS. Knowledge on the distribution of discontinuity zones in the Książ massif will enable designing a kinematic model of the rock massif behaviour during the effects of tectonic phenomena registered by the GL SRC instruments [30-32]. Moreover, the constructed high-resolution model may become a valuable element promoting such objects (natural or of anthropogenic origin) for visiting by tourists.

This work was financed by the Polish Statutory Research Grant no. 0402/0137/16 and Grant no. 0401/0128/17.

References

1. N.J. Rosser, D.N. Petley, M. Lim M., S.A. Dunning, R.J. Allison, Quarterly Journal of Engineering Geology and Hydrogeology, Vol. **38**, 363-375, DOI:10.1144/1470-9236/05-008 (2005)
2. J. Mah, C. Samsona, S.D. McKinnonb, D. Thibodeauc, International Journal of Rock Mechanics and Mining Sciences, Vol. **58**, 111-117, doi:10.1016/j.ijrmms.2012.08.001 (2013)
3. M. Dudziak M., *The secret of the Owl Mountains – guidebook* (JMK Publishing House, 1996, in Polish)
4. M. Aniszewski and P. Zagórski, 2006. *Underground World of the Owl Mountains – Riese – Książ Castle – Rüdiger* (Technol Publishing House, 2006, in Polish)
5. K. Grudzińska and D. Kasza, Hereditas Minariorum, vol. **1**, 113-123 (2014, in Polish)
6. CODGiK, Geodesic and Cartographic Documentation Center, Available online: http://www.codgik.gov.pl/, access date 5 November, 2017 (in Polish).
7. R. Marcinowski, J. Piotrowski, K. Piotrowska, *Dictionary of Polish lithostratigraphic units* (Polish Geological Institute, 2004, in Polish)
8. J. Oberc, *Tectonics. Sudeten Mountains and adjacent areas* (Geological Publishing House, Warszawa 1972, in Polish)
9. S.J. Porębski, Geol. Sudetica, **XVI**, No. 1, 101-192 (1981)
10. Z. Bełka and M. Narkiewicz, [In:] T. McCann (Ed.), *The Geology of Central Europe. Volume 1: Precambrian and palaeozoic* (The Geological Society of London, 383-410, 2008).
11. M. Kaczorowski and J. Wojewoda, Acta Geodynamica et Geomaterialia, **8**, No. 3 (163), 249-261 (2011)
12. D. Kasza, A. Kowalski, J. Wojewoda, M. Kaczorowski, E3S Web of Conferences 29, 00021 (2018)
13. M. Kaczorowski, Artificial Satellites, vol. **34**, no. 3, 171-191 (1999)
14. M. Kaczorowski, Artificial Satellites, vol. **34**, no. 3, 193-201 (1999)
15. M. Kaczorowski, Acta Geodynamica et Geomaterialia, **4**, No. 4 (148), 109-119 (2007)
16. M. Kaczorowski, Reports on Geodesy, **85**, 2, 79-86 (2008)
17. M. Kaczorowski, Acta Geodynamica et Geomaterialia, **6**, No. 3, 369-381 (2009)
18. M. Kaczorowski, 16th International Symposium on Earth Tides, 1–5 September 2008, Jena, Bulletin d'Information des Marées Terrestres (BIM), **144**, 11605-11613 (2009)
19. D. Kasza, M. Kaczorowski, R. Zdunek, R. Wronowski, Acta Geodynamica et Geomaterialia, **11**, No. 3 (175), 225-234, DOI:10.13168/AGG. 2014.0011 (2014)
20. R. Zdunek, M. Kaczorowski, D. Kasza, R. Wronowski, Acta Geodynamica et Geomaterialia **11**, No. 4 (176), 305-315, DOI:10.13168/AGG. 2014.0016 (2014)
21. M. Kaczorowski, P. Goluch, J. Kuchmister, K. Cmielewski, R. Zdunek, A. Borkowski, Acta

Geodynamica et Geomaterialia, **12**, No. 2 (178), 169-179, DOI:10.13168/AGG.2015.0012 (2015)

22. D. Kasza, *Recent tectonic activity in aspect of stability of the Książ Castle complex* (PhD Thesis, Wrocław University of Science and Technology, 2016)
23. MeshLab, official website http://meshlab.sourceforge.net/, access date 22.06.2015
24. M. Kazhdan, M. Bolitho, H. Hoppe, [In:] Symposium on Geometry Processing, Alla Sheffer and Konrad Polthier (Editors), The Eurographics Association, 61-70, DOI: 10.2312/SGP/SGP06/061-070 (2006)
25. M. Bolitho, M. Kazhdan, R. Burns, H. Hoppe, [In:] Advances in Visual Computing, Lecture Notes in Computer Science Volume **5875**, Springer, 678-689, DOI: 10.1007/978-3-642-10331-5_63 (2009)
26. M. Kazhdan and H. Hoppe, ACM Transactions on Graphics, Vol. **32**, No. 3, DOI: http://dx.doi.org/10.1145/2487228.2487237 (2013)
27. M. Bolitho, M. Kazhdan, R. Burns, H. Hoppe, [In:] Symposium on Geometry Processing, 69-78 (2007)
28. M. Corsini, P. Cignoni, R. Scopigno, IEEE Transaction on Visualization and Computer Graphics, Volume **18**, Number 6, 914-924 (2012)
29. M.S. Muhammad, M.T. Mahamood, C. Tae-Sun, IEEE International Conference on Acoustics, Speech and Signal Processing, 753-756, DOI: ieeecomputersociety.org/10.1109/ICASSP.2009.4959693 (2009)
30. M. Kaczorowski, Acta Geodynamica et Geomaterialia, **2**, No. 3 (139), 1-7 (2005)
31. M. Kaczorowski, Acta Geodynamica et Geomaterialia, **3**, No. 3 (143), 79-84 (2006)
32. M. Kaczorowski, [In:] Earthquake Source Asymmetry, Structural Media and Rotation Effects, Springer-Verlag, Berlin, 493-520 (2006)

Three-dimensional model of the Rzeczka object (the Riese Complex, Sowie Range) as an enhancement of the comprehensive geotourism potential of an underground tourist route

Katarzyna Kądziołka[1], *Damian* Kasza[1,*], and *Wojciech J.* Milczarek[1]

[1]Wrocław University of Science and Technology, Faculty of Geoengineering, Mining and Geology, 27 Wyb. Wyspiańskiego St., 50-370 Wrocław, Poland

Abstract. The dynamic development of measurement technology based on the use of laser scanners allows for fast transfer of information, related to terrestrial data, to a virtual environment and enables the creation of three-dimensional models. Because of the growing demand for underground areas inventory, especially historic areas, the creation of 3D numerical models increases in popularity. It allows to increase the geotouristic potential of a particular area and to use solutions in support of the safety of tourists as well as the safety of area itself. The development of the terrestrial laser scanning technique is accompanied simultaneously by the development of means of digital data processing that was obtained in the form of point clouds. Those means use the commercial software as well as software available in the *open source* system. Both computing environments use algorithms enabling the processing of huge collections of points in automatized mode by using a number of open-access algorithms either freeware or dedicated. The methodology of numerical models development, described in this article, has been presented on the example of underground object "Rzeczka" ("Riese" Complex) located at the Sowie Range (SW Poland).

1 Introduction

Currently, the terrestrial laser scanning is the fastest growing technique used for obtaining terrestrial data for closed areas. The advantages of this technique include high-speed data acquisition, high precision of measurement, obtaining information about the physical characteristics of examined/measured areas and the independence from lighting conditions, among many others [1].

The emitter of laser beam built into the scanner (own source of light) makes this measuring instrument perfect for underground areas inventory. However, the measurement itself is not all. Data processing and creation of numerical models are significant problems. Currently, algorithms used for the reconstruction of a surface out of point clouds allow for working with multimillion collections of data with a different degree of accuracy. The access to these collections can be obtained from the level of *open source* as well as commercial type of software.

In the era of growing popularity of underground complexes as geotouristic areas, the promotion of this type of areas becomes very important. The numerical three-dimensional models contain information about the present state of the area and can easily be used as an advertised product in the form of interactive visualizations and animations.

2 Study area

"Rzeczka" is one of seven currently discovered underground objects of a secret Nazi mining-construction project done under the cryptonym "Riese". The project envisaged building a complex of interconnected underground excavations and ground infrastructure which most likely were supposed to function as factories and laboratories [2, 3].

A complex of underground tunnels of "Rzeczka" has been excavated in the rock massif of Ostra Mountain in the Sowie Range (Lower Silesia, SW Poland). Three adits leading into the area are located at the East Mountainside of the above-named massif at the heights from 557 to 559 meters above sea level [2]. The available tourist route incorporates the whore area of underground excavations and is about 500 m in length whereas area's cubature is 14 000 m^3 (Fig. 1).

* Corresponding author: damian.kasza@pwr.edu.pl

 https://doi.org/10.1051/e3sconf/20185500015

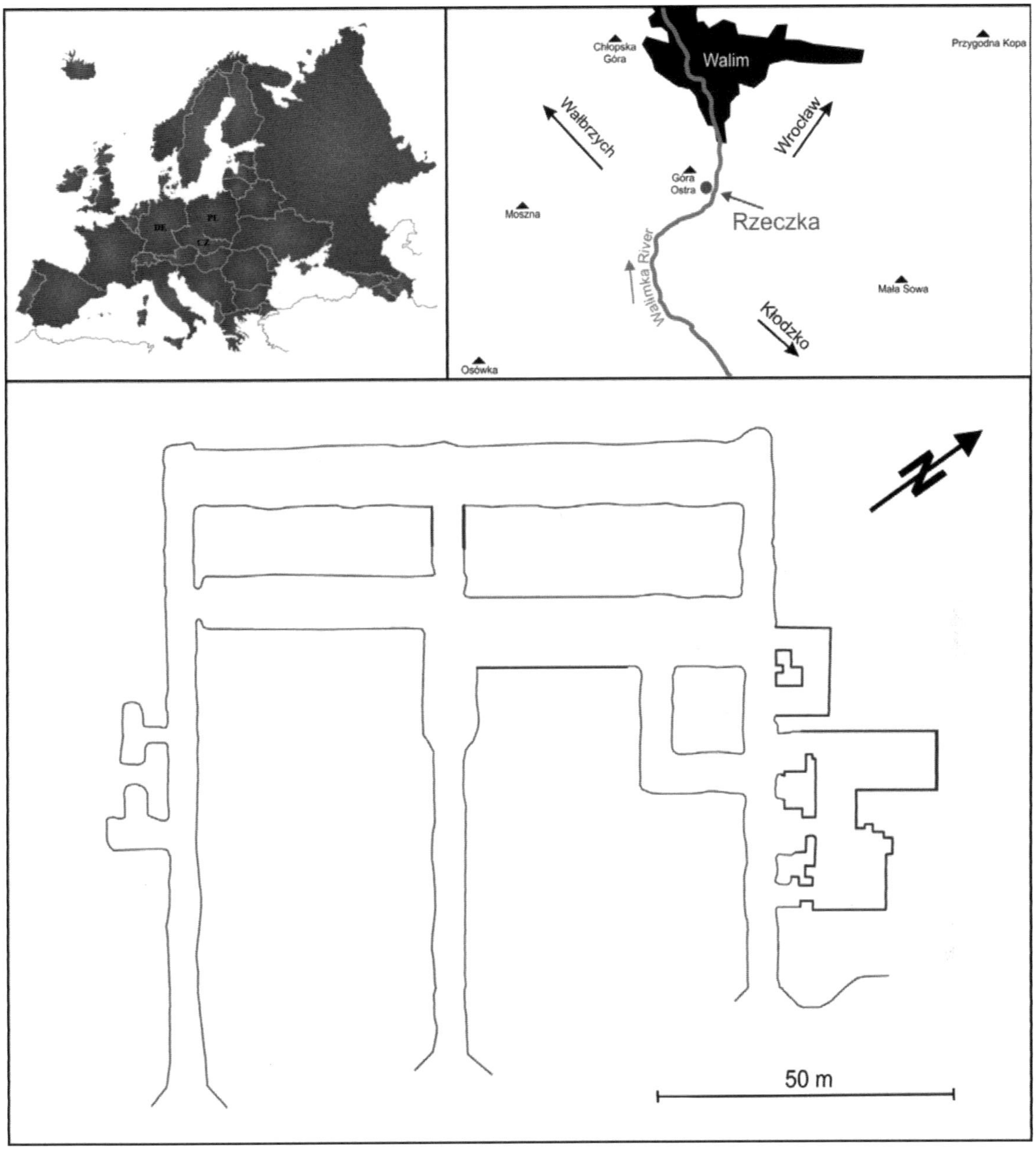

Fig. 1. Localization of "Rzeczka" underground complex and the area's plan.

3 Terrestrial laser scanning technology

Terrestrial laser scanning, also known as LiDAR (*Light Detection and Ranging*), is an active remote sensing system which uses instruments that emit electromagnetic waves in a form of the laser beam. Scanning is relatively new and ever-improving technique that involves collecting terrestrial data of examined areas and obtaining the location of these areas (points) in terms of the scanner. It allows for the fast acquirement of dense point models, in which every point is assigned not only to *xyz* coordinates but also to an additional parameter which marks an intensity value of reflected wave [1].

The allocation of coordinates *xyz* of a particular measurement point in the scanner's local system is executed by measuring the distance, vertical angle and horizontal direction. Basing on the knowledge of distances and directions of the point, the coordinate increments of the particular point are determined in terms of scanner's location and elevation (Fig. 2):

$$\Delta x = s \times \cos\alpha \quad (1)$$

$$\Delta y = s \times \sin\alpha \quad (2)$$

$$\Delta z = s \times \cos\beta \quad (3)$$

where s denotes the distance between the scanned point and scanner, α denotes the value of the horizontal angle, β denotes the vertical angle.

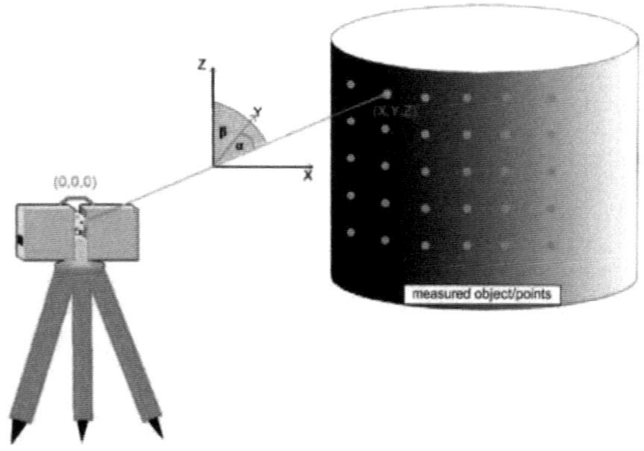

Fig. 2. The graphical representation of the essence of measurement taken by the laser scanner [4].

During the inventory works in "Rzeczka", the Leica's ScanStation C10 scanner has been used. Measurement stations were positioned not more than 5 meters away from each other in order to assure covering the whole area. The scanning resolution was 5x5 cm per 50 meters of the distance. The HDS targets with a diameter of 6" were used as the reference points inside the underground excavations to connect individual measurement stations of the scanner. The final point cloud, created during the post-processing in the Leica Cyclone environment, has been linked to coordinate system PL2000 zone 7 by using 3 points which were marked in front of adits and measured with the use of RTK GPS technique.

4 The analysis of the results

The obtained data set consisted of nearly 120 million points. The point cloud was pre-treated in order to eliminate noise – incorrectly measured points. Some of it (that represented scanned operators' silhouettes, support stands with reference targets, etc.), has been removed manually. After that, the whole point cloud was filtrated by using the *Noise filter* tool, available in the Cloud Compare environment [5]. Ultimately – for more effective work related to mesh type surface reconstruction, the cloud was simplificated by the use of *Poisson-disk sampling* tool available in MeshLab application program [6]. This operation allowed for the unification of set by determining the minimal distance between points within the given value of 10 cm. Thereby, the point cloud, after the pre-processing phase, consisted of nearly 14 million points and was prepared for the second phase of works – the creation of mesh type surface structure with the use of open source type of software – MeshLab and commercial type of software – SketchUp.

4.1 Surface reconstruction of mesh type in MeshLab application program

MeshLab's environment has a few built-in algorithms allowing for generation of mesh type structure model based on data derived from laser scanning. For conducting a series of tests, the authors of this paper decided on picking 3 algorithms: *Marching Cubes*, *Ball Pivoting* and *Poisson Surface Reconstruction*.

4.1.1 Marching Cubes

This algorithm was designed by Lorensen and Cline in 1987 [7]. It works by creating a remote scalar field of a triangle mesh that approximates the equipotential surface. The surface is created by connecting 26 points that have equal value. In the Marching Cubes algorithm the scalar fields, built out of eight vertices that create a cube, are being analysed (Fig. 3) [8]. Each vertex is assigned with a value of 0 (when a vertex is located outside of the constructed surface) or 1 (when a vertex is located inside or on the wall of the presented surface).

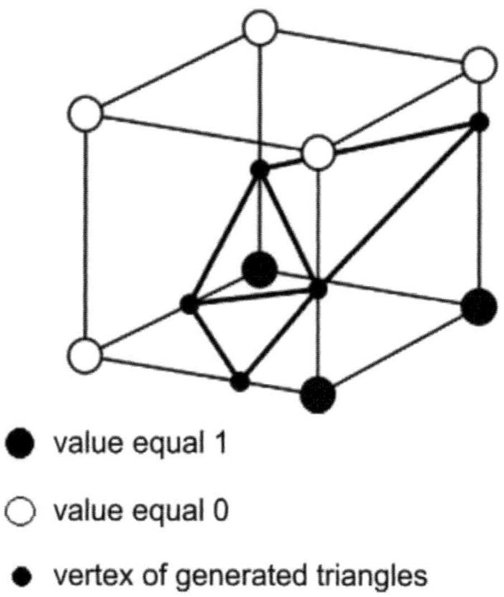

Fig. 3. The graphical representation of a structure of triangle mesh with the use of the Marching Cubes algorithm [8].

The process of generating the grid of triangles is based on counting up all the vertices placed on the edges of constructed cubes and connecting their intersections with cubes' surfaces to create triangles. The resolution of generated mesh, by the Marching Cubes algorithm, can be regulated by setting the number of cubes. The bigger the number of cubes, the smaller the dimension of each cube and consequently the greater accuracy of the model's projection (more details are included). However, each increase in number of cubes increases the time required for generating the triangle mesh that represents the model's surface [7].

4.1.2 Ball Pivoting

The creation of triangle mesh, with the use of Ball Pivoting algorithm, starts with determining the radius of the sphere that rotates around all the edges of created triangles. The moment when the sphere touches another point, which is contained in the point cloud, begins the appointing of a new triangle, made out of two points of the previous triangle that the sphere rotated around, and appointing of the third point which has just been encountered by the sphere. In this way, consecutive triangles come into being and create triangle mesh of the model. The rotating sphere generates further parts of the triangle mesh as long as its surface detects other points of the cloud (Fig. 4). In less frequent positioning of points or in areas of insuficient data, the generated model will have multiple gaps that entail an inaccurate projection of the area. The algorithm is suitable mainly for densely and evenly distributed points [9].

The precision of area's projection with the use of *Ball Pivoting* method depends on accurate choosing of the rotating sphere's radius. If the value of the radius is too small, there is a possibility of skipping many details as well as creating multiple gaps in the model [9].

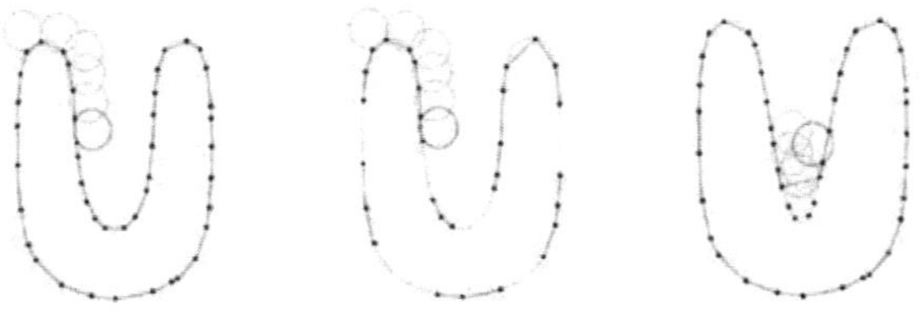

Fig. 4. The graphical representation of a structure of triangle mesh with the use of Ball Pivoting algorithm [9].

4.1.3 Poisson Surface Reconstruction algorithm

The *Poisson Surface Reconstruction* algorithm was designed by Kazhdan, Bolitho and Hoppe [10-13]. Oftentimes, this algorithm is implemented in programs that are used for developing point clouds. It is used particularly during the process of automatized creation of irregular triangle mesh TIN. The *Poisson Surface Reconstruction* algorithm is one of the most frequently applied one, among those that are being used for creating surfaces out of large collections of points, due to it's operating mode - it is a global method that employes the whole data set during the computational process without dividing it into smaller sample points for the purpose of approximating the modeled surface. According to this fact, the algorithm is immune to undesirable data distortion (e.g. gaps in point clouds).

The algorithm's work is based on the solution of the Poisson's equation. For the purpose of application of this algorithm, it is required beforehand to calculate normal vectors of all points of the formulated set. The first stage of the algorithm's work is to define the gradient of the particular set function and to determine its' simplified characteristic function. The next stage consists in approximating the gradient of the scalar field. For optimal alignment of the data set, the aligning filter based on Gauss equation is applied. The appointed in this way irregular triangle mesh TIN, which forms the surface of the modeled area, has the following features:

- huge detailedness of the projection of model's edges
- high immunity to the formation of distortion
- during the computational process an octree method is employed (lower computational resources load)
- a complete, so-called "closure" of the model (leaves no 'holes' in the model) [10].

4.2 Surface reconstruction of mesh type in SketchUp application program

The SketchUp software is a commercial environment which allows for 3D modeling, point clouds servicing, modeling buildings and other engineering structures. The software extended with the Undet plug-in allows for importing the point cloud. It also contains many tools allowing for efficient management and visualization of data. Undet enables generating and editing the mesh type surfaces. The Undet 2017 version of the program has an option for automatized detection of surfaces and construction of a grid of irregular triangles on the pointed area - it allows for determining mesh's parameters: size and curvature tolerance [14].

5 Results and discussion

The conducted tests for generating surface's structure of mesh type, for the chosen fragment of excavation by using 3 different algorithms available in the MeshLab software, brought diverse results. The Marching Cubes, as well as the Ball Pivoting algorithm, projected the model of underground excavations in an unsatisfactory degree. The model had multiple gaps (so-called holes, outgrowth, and other artifacts; Fig. 5A, B). In the microscopic evaluation of created models, the most detailed model has been generated by the *Poisson Surface Reconstruction* algorithm (Fig. 5C).

The test results presented in Fig. 5 made the choice clear and the *Poisson Surface Reconstruction* (PSR) algorithm has been used for constructing the model of "Rzeczka". In the first step, for prepared data set in the point cloud form (cf. Section 4.), normal vectors, which are necessary for the algorithm's work, were generated. Unfortunately, scanner's and software's producer does not provide the option for exporting information about normal vectors directly along with the collection of points. Normal vectors were generated with the use of *compute normals for point sets* tool. The process of calculating these vectors consists in determining the approximate local surface by using the location of points and their nearest 'neighbours'. The normal, in the specified point, vectors are perpendicular to the specified surface whereas the surface is set by the estimated triangulation method with the use of neighbouring points [6].

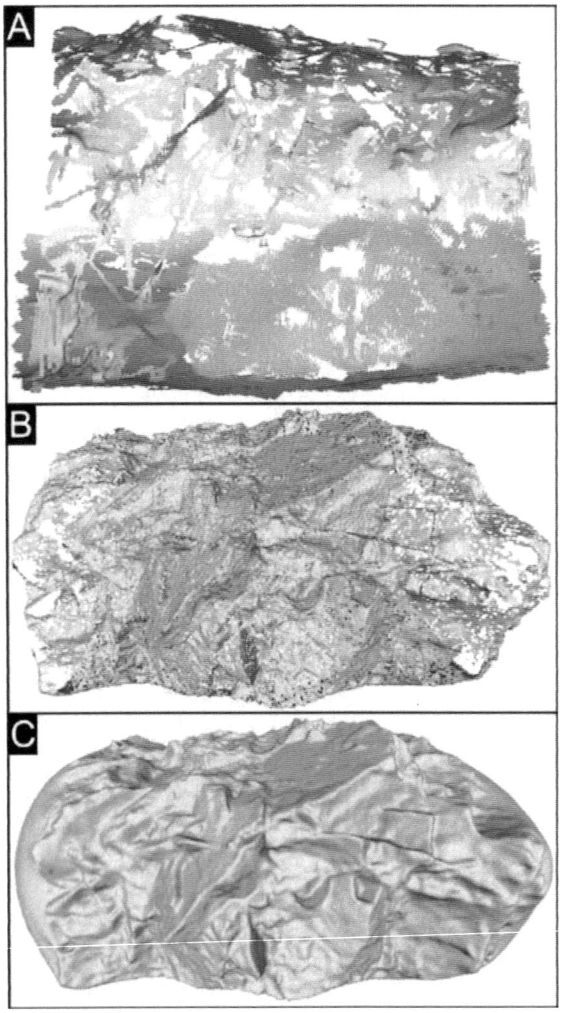

Fig. 5. The test results for mesh type surface creation with the use of Marching Cubes (A), Ball Pivoting (B) and Poisson Surface Reconstruction (C) algorithms.

The second step was the beginning of testing the parameters that create triangle mesh – mesh type structure with the use of the PSR algorithm. During the process, vertices, edges and surfaces representing the model are generated. These features are tangential only along the edges [15]. The most important parameter, which is responsible for the accuracy of area's projection, is the depth parameter (*octree depth*). Its' value determines the complexity of generated mesh and consequently the detailedness of the model. This parameter has a great influence on the run-time of the computational process. The higher its' value, the more time-consuming computational process. For the purposes of comparison, the model was generated with the depth being set to 8 (Fig. 6A) and 15 (Fig. 6B). The number of generated triangles depending on the parameter's value was presented in Table 1.

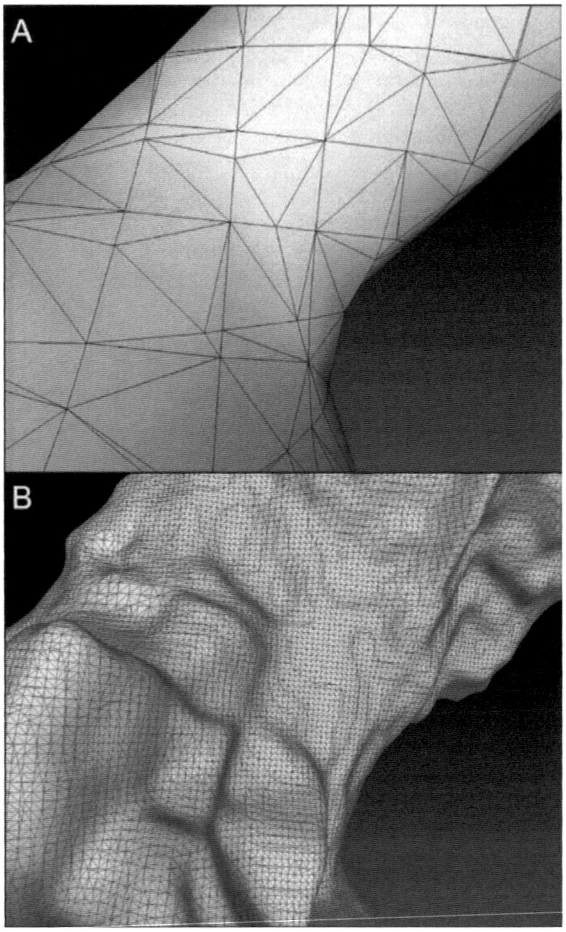

Fig. 6. A fragment of the model generated with the use of Poisson Surface Reconstruction algorithm: depth parameter set to 8 (A) and 15 (B).

Table 1. The comparison of the number of generated triangles depending on the value of depth parameter.

Octree depth	Number of triangles
8	60 568
15	17 103 642

Based on the presented comparison of generated models, one may conclude that there is a major difference in the accuracy of projecting model's details. The model with depth value set to 8 is clearly less accurate. Sharp edges of the underground excavation, which has complex geometry, were overly smoothed, therefore, the model does not represent the actual image of the area. The model's triangle mesh with depth value set to 15 is significantly more complex what makes the projection of the geometry of excavations more accurate. The details of the projected surface and its' sharp edges were preserved. The authors did not conclude the enhancement of model's detailedness along with a further increase in depth parameter's value. The final model is presented in Fig. 7.

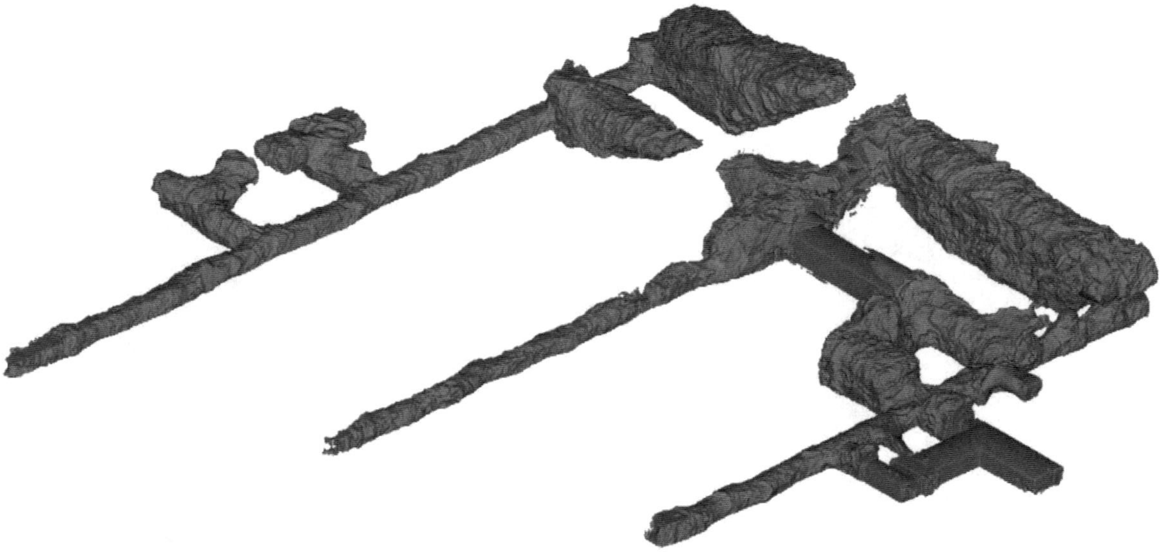

Fig. 7. The numerical model of the underground excavations of "Rzeczka" built in the MeshLab environment.

The three-dimensional model of "Rzeczka", in SketchUp's environment, was developed in Undet extension with the use of *Feature extraction* tool. In accordance with the description [14], this tool is suitable for all types of surfaces - buildings, engineering structures, roads, ground's surfaces, tunnels or caverns. In the presented case, this option generates surfaces in a specified (given) area of point clouds in a highly inaccurate manner, probably due to the complex geometry of the excavation (Fig. 8A). In the case of enclosed elements of the area, e.g. guardroom (simple geometric shapes), the surfaces were adequately designated and consequently the generated TIN mesh accurately represents surfaces/planes of these rooms (Fig. 8B).

The authors concluded that the point clouds conversion of the scanned area, which has complex geometry, into the mesh of irregular triangles done in the Undet for SketchUp program, requires constant evaluation of correctness on the user's part, i.e. determining the placement of the surface that was created with the use of *Feature Extraction* option, creating the mesh surface with the use of *Convert to Mesh* option and manual editing of the mesh: deletion of mistakenly generated triangles or change of surface's orientation that was proposed in the program. During the transformation of the designated area into the mesh structure, it is necessary to set the parameter of surface's curvature tolerance to maximum - in that, it is possible to project irregular and complex geometry of excavations, and to generate the numerical model (Fig. 9).

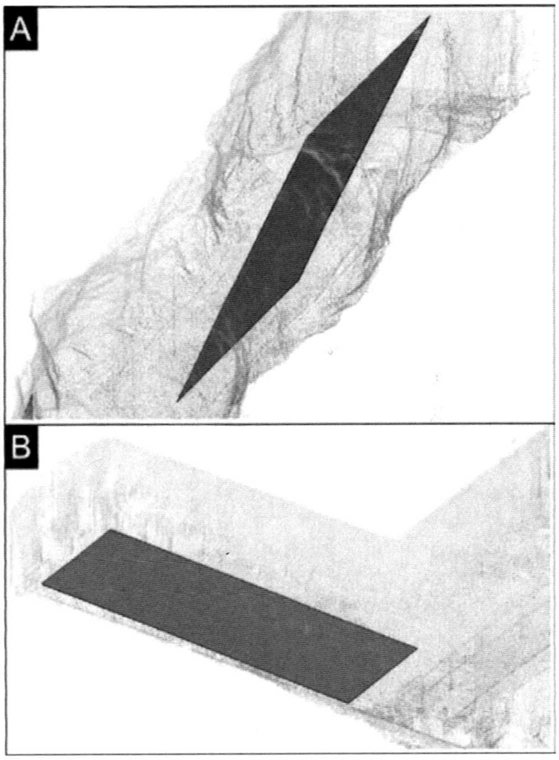

Fig. 8. Example of employing Feature extraction tool: incorrectly designated surface (A) and correctly designated surface (B).

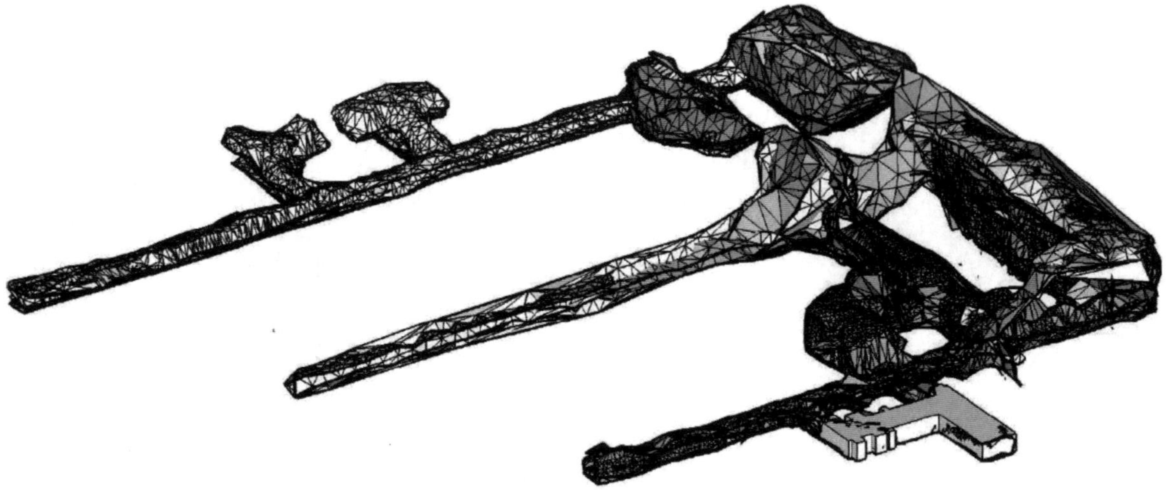

Fig. 9. The model of "Rzeczka" generated with the use of Undet for SketchUp plug-in.

6 Conclusions

The foregoing layouts, that present the areas of the Riese complex, were created with the use of classical methods, i.e. tacheometric surveying and levelling surveying, which enabled the creation of various plans of the objects, projections or maps. The terrestrial data extraction, done with the terrestrial laser scanning method, allowed for creating the numerical representation of "Rzeczka" in the form of three-dimensional model which presents the geometry of the whole area in detail. The created model enables further processing for the purpose of conducting subsequent analyses regarding the construction or dimensions of the complex as well as creating derivative products on the model's basis, i.e. maps or profiles.

The three-dimensional reconstruction of the surface can be done in multiple graphical environments - open source as well as commercial type. The MeshLab program enables creating the reconstruction in a highly effective manner - due to the implemented algorithms, the surface reconstruction process has been done with a huge precision in a short period of time. The reconstruction of the surface that has complex geometry in the Undet for Sketchup programme is a very time-consuming process and the generated TIN mesh does not present the geometrical details. In the case of generating models of regular shapes, e.g. guardroom located at the undergrounds of "Rzeczka" object, it needs to be noted that the Undet program is good with recognising surfaces of noncomplex geometrical shapes.

The model created in the MeshLab environment, due to the possibility of further processing and capability of exporting it in various formats, can also be used as a promotion tool, e.g. in a traditional way (various types of printouts) or in a digital way as an animation or interactive model available on the website.

This work was financed by the Polish Statutory Research Grant no. 0402/0137/16 and Grant no. 0401/0128/17.

References

1. G.L. Heritage and A.R.G. Large, *Laser scanning for the Environmental Sciences* (Wiley-Blackwell, 2009)
2. P. Kruszyński, *Podziemia w Górach Sowich i Zamku Książ* (Muzeum Gross-Rosen, Wałbrzych, 2004)
3. J. Rostowski, *Podziemia III Rzeszy. Tajemnice Książa, Wałbrzycha i Szczawna-Zdroju* (Dom wydawniczy Rebis, Poznań, 2010)
4. R. Staiger, *Terrestrial Laser Scanning Technology, Systems and Applications* (2nd FIG Regional Conference, Marakkech, 2003)
5. CloudCompare, official website www.cloudcompare.org, access date 22.10.2017
6. MeshLab, official website http://meshlab.sourceforge.net/, access date 22.10.2017
7. W.E. Lorensen and H.E. Cline, Computer Graphics, **XXI(4)**, 163-169 (1987)
8. C. Montani, R. Scateni, R. Scopigno, The Visual Computer, **10** (6), 353-355 (1994)
9. F. Bernardini, J. Mittleman, H. Rushmeier, C. Silva, G. Taubin, IEEE Transactions on Visualization and Computer Graphics, **5** (1999)
10. M. Kazhdan, M. Bolitho, H. Hoppe, [In:] Symposium on Geometry Processing, Alla Sheffer and Konrad Polthier (Editors), The Eurographics Association, 61-70, DOI: 10.2312/SGP/SGP06/061-070 (2006)
11. M. Bolitho, M. Kazhdan, R. Burns, H. Hoppe, [In:] Symposium on Geometry Processing, 69-78 (2007)
12. M. Bolitho, M. Kazhdan, R. Burns, H. Hoppe, [In:] Advances in Visual Computing, Lecture Notes in Computer Science Volume **5875**, Springer, 678-689, DOI: 10.1007/978-3-642-10331-5_63 (2009)

13. M. Kazhdan and H. Hoppe, ACM Transactions on Graphics, Vol. **32**, No. 3, DOI: http://dx.doi.org/10.1145/2487228.2487237 (2013)
14. Undet for SketchUp, official website www.undet.com, access date 22.10.2017
15. F. Remondino, International Archives of Photogrammetry, Remote Sensing and Spatial Information Sciences, Vol. **XXXIV-5/W10** (2003)

AUTHOR INDEX

Bac-Bronowicz Joanna 17
Blachowski Jan 10
Bujarkiewicz Adam 76
Gradka Robert 81
Gruchlik Piotr 51
Grzesiak Karolina 38
Herkt Paulina 10
Kaczorowski Marek 1
Kadziolka Katarzyna 102
Kasza Damian 1, 95, 102
Kowalski Andrzej 51
Kwinta Andrzej 88
Milczarek Wojciech J 25, 38, 102
Mrówczynska Maria 61, 76
Odziemczyk Waldemar 66
Owczarz Karolina 31
Sudo Ewa 47
Sztubecki Jacek 61, 76
Wajs Jaroslaw 25
Wazydrag Karolina 88
Wojciechowska Gabriela 17
Wronowski Roman 1
Zdunek Ryszard 1
Zygmunt Mariusz 88